Paid

Infrastructures Series

edited by Geoffrey C. Bowker and Paul N. Edwards

Paid

Tales of **Dongles,**

Checks, and **Other**

Money Stuff

edited by **Bill Maurer** and **Lana Swartz**

foreword by **Bruce Sterling**

THE MIT PRESS
CAMBRIDGE, MASSACHUSETTS
LONDON, ENGLAND

© 2017 Massachusetts Institute of Technology

This book was set in Archer and DIN by the MIT Press. Printed and bound in the United States of America.

Library of Congress Cataloging-in-Publication Data

Names: Maurer, Bill, 1968- editor. | Swartz, Lana.
Title: Paid : tales of dongles, checks, and other money stuff / edited by Bill Maurer and Lana Swartz ; foreword by Bruce Sterling.
Description: Cambridge, MA : MIT Press, [2017] | Series: Infrastructures | Includes bibliographical references and index.
Identifiers: LCCN 2016028477 | ISBN 9780262035750 (hardcover : alk. paper)
Subjects: LCSH: Money--History. | Electronic funds transfers. | Automated tellers. | Telematics.
Classification: LCC HG231 .M58655 2017 | DDC 332.4--dc23 LC record available at https://lccn.loc.gov/2016028477

10   9   8   7   6   5   4   3   2   1

Contents

Foreword
Dead Money

BRUCE STERLING

I was charmed by this book. It's chock-full of wonder and sadness.

I'm a novelist, but also an amateur historian of media. In my historical studies, I look for data with page-turning qualities, something eye-catching, marvelous, and maybe grotesque. Something that offers a high "Cahill Factor," with the quirky and scarcely credible qualities of the legendary Thaddeus Cahill's "Telharmonium."

Cahill's Telharmonium, you see, was once a gigantic, sophisticated electronic music production and distribution system. It debuted in 1906. Mark Twain was a happy subscriber to Mr. Cahill's commercial music-streaming service. The Telharmonium was the technical state of the art for Gilded Age Manhattan. Millions of dollars were wasted on this grandiose device. A mastodon in a tar pit couldn't have died a more horrid and lingering death than this brilliant yet utterly doomed network machine. I've never yet written a work of fiction about Cahill's Telharmonium, but since I know something about it, I can write books with heartfelt, melancholy titles such as *Gothic High-Tech*.

The Cahill Telharmonium, however, for all its merits, was never nearly so vast, so all encompassing, so ambitious, so horrible, so dinosaurian as obsolete money systems. This book is all about archaic systems of financial exchange.

It's not a book about the fine hobby of numismatics, for metal coins, as physical objects, are romantic and pretty. It's about the debris that are paper checks, bills, tickets, stubs, files, records, logs, accounts, and receipts, or magnetic stripe cards, dongles, Minitel units, and defunct ATM Bancomats—a huge variety of money management technologies, but, well, they're all trash. Nobody mulls them over with the severe joy that collectors experience with their Roman imperial coinage and obscure French postage stamps. This dead financial hardware is disgusting rubbish, even pollution.

They were once of severe and painful value—the burden of a mortgage can kill you; a check that bounces is a lasting humiliation—but their worth on earth is brief, while their condition as junk is lastingly abject.

I surmise that this is why so many of the authors in this book feel a need to apologize for their keen interest in the topics they so entertainingly describe. This book is chockablock with technical wonderment. Who knew, for instance, that Benjamin Franklin printed fallen leaves into colonial American money because all-natural leaves are so hard for human beings to counterfeit? Who knew that the wife of Kwame Nkrumah kept a secret stash of Egyptian cash, robbed from her palace in an African coup d'état? But behind these bright sparks of historic erudition is a lasting air of mortal weltschmerz.

Why have we done such awful things to ourselves, just for our all-too-mortal systems of money? Take the Native Americans of California, for instance. These fortunate people were living in an area of nigh-utopian natural wealth and beauty, so it's startling, and also depressing, to learn that these early inhabitants invested brutal effort and weird ingenuity in scraping and grinding coin-like tokens from pretty Californian seashells. Not only were these wampum-like strings of "shell beads" of critical importance to their own hunter-gatherer society, but they seemed to have no trouble at all exporting this system of value to everyone they could reach. They were the Silicon Valley of seashells as money.

People believe in money. But it just doesn't last.

To judge by our modern ingenuity in storing money, shipping money, and repeatedly wrecking our society with vicious financial panics, nobody's ever believed in money quite like we moderns do. What was once merely the root of all evil is now the root of our every whirring data packet. It's a grim tale, and yet this fine book conveys a heartening sense of memento mori. This too will pass. All too soon, the dismal banking systems that pester us nearly to madness will be as corny and archaic as the French Minitel, whose national saga, deftly touched on herein, is even sadder than an Édith Piaf song.

Dead media can be extremely funny; silent film comedies are often wondrously hilarious. Money, by contrast, is never amusing. It's extremely rare—unheard of, in fact—to find a coin or paper bill whose designers made a joke of it, or took the opportunity to say something witty to millions of users. Tombstones are more lighthearted than money is.

Yet there is one exception, as detailed in this book: the astonishing tale of the Dogecoin, an out-of-control Internet crypto coin that started as a wry meme joke and somehow became a form of money. Given that the Dogecoin was more or less money, though, a huckster and fraudster promptly arrived to ruin that digital party. Why? How? Read on.

Money is a distributed network computer whose circulation calculates value. That is its purpose. That is, unless you're ancient and Inkan, in which case you've got a system of knotted and colored cords of llama yarn that is ideal for forcing preliterate people to work for their overlords. Money is always a system of abstracted exploitation. It's like a language whose only possible vocabulary is who does what for whom.

That's why people don't love money. If you've spent your life writing diaries, you're hard put to burn them, even if your death is at hand and they're full of indiscretions. But if you've spent your life perched on Bob Cratchit's clerking stool writing the double-entry books for Scrooge, everybody, including you, is secretly overjoyed when Scrooge and Marley's warehouse falls to the purifying flames of history.

I'm a novelist myself, so I know that Charles Dickens was getting away with literary murder in that parable about the innate cruelty of money; he was utilizing ghosts, Christmas, and the crippled kid all to pretend that the people of Victorian Britain weren't the planet's financial overlords at the time of his own writing. Scrooge, Cratchit, and even Dickens were all making out like bandits at the cost of the rest of us. Everybody's guilty; no one's hands are clean. Why did Dickens write that story? Why does it last? Because money's like that, and literature isn't.

They were sinners, but so are we; the big difference is the Victorians lugged gold bullion in big canvas sacks while we rely on buggy, cranky, digital Rube Goldberg schemes that will soon be as dead as eight-track tapes and floppy disks. We've got authenticators, we've got connectors—but mostly we've got cheesy, reeking, morally rotten hacks.

Who will look with tenderness and understanding on our financial kludges, on the havoc we wreaked for objects and services with the life span of hamsters? The Ghost of Christmas Future, maybe. And the successors of the insightful people in this book.

Acknowledgments

We would like to thank Tom Boellstorff and Kevin Driscoll for their support and encouragement while we were completing this volume. We would also like to thank Jenny Fan, Julio Rodriguez, the Institute for Money, Technology, and Financial Inclusion (IMTFI) at UC Irvine, and the (former) Intel Science and Technology Center for Social Computing at UC Irvine for supporting the initial convening at which this volume was first envisioned. We thank Robert J. Kett for his curatorial work at IMTFI, which directly inspired this book, and for his help on our first "payments party." We also thank the other participants of that convening: Mic Bowman, Alejandro Komai, Juliet Levy, Henry Lichstein, Daniel Littman, Erik Moga, Mark Moore, Donald J. Patterson, Katherine Porter, Gregory Waymire, and Irving Wladawsky-Berger.

Lana would like to thank Manuel Castells for his generous intellectual and material support. She received funding through the Wallis Annenberg Chair in Communication, Technology, and Society fellowship and the Balzan Foundation during work on this project. She would also like to thank Jennifer Chayes, Nancy Baym, Mary L. Gray, Tarleton Gillespie, Sarah Brayne, and Microsoft Research New England.

Maurer's research on payment systems and technologies has been supported by the US National Science Foundation (SES 0960423 and SES 1455859). Any opinions, findings, and conclusions or recommendations expressed in this material are those of the author(s) and do not necessarily reflect the views of the National Science Foundation.

We are very grateful to everyone at the MIT Press for their work on this book. Series editors Geof Bowker and Paul Edwards saw merit in this book from the beginning. Katie Helke shepherded it through the next steps. Anonymous reviewers provided excellent helpful guidance and criticism for which the book is the better. We would also like to thank Margy Avery, Deborah Cantor-Adams, Susan Clark, Erin Hasley, Justin Kehoe, and Cindy Milstein.

This book is dedicated to Cooper and Rufus, who have never had to pay for a thing.

Introduction
Curating Transactional Things

BILL MAURER AND LANA SWARTZ

Consider this book a catalog for an exhibition that never happened. Each contribution takes up an ephemeral object connected to the act of payment, a transactional thing that unlike metal coins or paper banknotes would rarely make it into a display case—or, in some cases, even be impossible to curate. Money is the most obvious transactional thing. It is used for billions of transactions every minute, the sum total of which we call an "economy." Museums have money galleries, displaying coins, notes, and some of the other objects that people have used historically to conduct transactions—pretty things like beads, shells, stones, and metal ingots. Such items have been displaced by coins and cash, and commentators today predict the imminent demise of those transactional things, too, heralding a cashless future in the face of digital technologies of payment. Supposedly, the pretty things will get replaced by pure data.

Increasingly, however, monetary transactions have begun to require even more things. There are payment cards, mobile phones, electronic point-of-sale terminals, networks of wire, and webs of radio signals. Money-like things, too, have started to multiply. More and more, people are using frequent-flier miles, coupons, cryptographic currencies, and sharing economy platforms to conduct transactions. As sociologist Viviana Zelizer puts it, "money multiplies," and how.

These payment objects and some of their ancient antecedents are the subject of this book—the transactional things that are often forgotten, ignored, or operate in the background, and the social, artistic, and political practices that sometimes bring them to light. This book is one such practice. It brings together writers from archaeology, history, technology studies, industry, law, and computer science, each of whom considers an object, turns it around on the page, and reflects on its place in the archive of human exchange.

The era of cash and coin, of tangible, physical objects serving to settle transactions, is relatively speaking a historical anomaly, especially seen from the point of view of ten thousand years of recorded human civilization. Archaeologists and historians of the ancient Near East have shown that money of account, recorded in transactional records, long predated the minting of coin or other tangible objects used as a universal equivalent for exchange. In the beginning was not the coin but rather the receipt.

So coins and banknotes are not the only transactional things. And receipts do not make themselves: there are whole technological and institutional apparatuses around the recording of transactions. Anthropologists have long talked about "gift economies," which use other kinds of objects to record the transactions that tie people and groups together.

A great deal of attention has been paid to the thingness of traditional money. The US Mint calls coin collecting "the king of hobbies and the hobby of kings." Most major museums feature a collection of gold coins, curious banknotes, counterfeits, and ethnographic curios of shell, stone, beads, and bones all taken as tokens of value. There is the sense that transactional things, in the form of money, connect us to the past and should be preserved.

But with rare exception, there are few people or institutions interested in preserving the stuff of new transactional things. Where are the charge-a-plates, the zip-zap machines, and mobile phone dongles? Today's payment artifacts—along with their associated merchant guidebooks, authorization checklists, wires, and cables—are mostly regarded as trash. This is true even though pundits are perpetually proclaiming the "death of cash," and even though the "cashless society" has been part of an idealized future for over a century.

One reason for this is because new transactional things don't really seem like things. How, exactly, would a museum curator collect and display Bitcoin, the cryptographic currency (the British Museum has tried)? While some professors, entrepreneurs, and artists have made physical Bitcoins, these are mostly novelty items. Bitcoin "lives" in the blockchain, a dynamic transaction record produced in concert by a network of decentralized computers. Or what about the Automated Clearing House, the infrastructure responsible for much of today's direct deposit and direct bill payment (though at least one money museum, at the Atlanta Federal Reserve, sought to curate it, owing to its significance for that regional branch of the United States' central bank)?

Another reason why contemporary transactional things are rarely paid attention to is because it is part of their job to be invisible. The Square dongles that Scott Mainwaring describes in chapter 1 are given away for free, adding to their "frictionless" aesthetics. Like many critical infrastructures, most users only notice them when they are broken. When a point-of-sale terminal breaks, a merchant either throws it away or returns it to the third-party payment company that provided it, who will probably throw it away as well. The online market place eBay is littered with such wreckage. So are electronic waste collection facilities.

In addition, the payments industry is little understood by those not directly involved in it, and there are many, many pieces of the payments puzzle. Even the basic card networks like Visa and MasterCard depend on issuing and acquiring banks, payment processors, point-of-sale devices, independent sales organizations, manufacturers of plastic cards, "J-hook" displays from which gift cards are suspended in grocery stores, wireless networking services, and chip and device manufacturers. At the level of infrastructure, many services depend on the "rails" and "pipes" (to use industry parlance) of other services: Western Union "rides the rails" of First Data; PayPal uses the federally mandated Automated Clearing House.

We want to encourage readers to start pulling these things apart, to look at their inner workings. Looking under the hood, into the machinery, of everyday transactions, seeing all the different players in a system described by participants as an "ecology," provides new insight into what we shorthand as "the economy." This comes at an important political moment. The global financial crisis, increasing income inequality, and the spread of mobile computing and world of connected digital devices all together have opened up new questions about what money is, how the act of payment takes place, and whether it is time for a reconsideration of value itself. There is an explosion of interest in alternative currencies, new value formations based on reputation or trust in online networks instead of traditional forms of labor or investment, and the rise of the so-called sharing or peer economy. A lot of people are spending a lot of time (and good-old traditional forms of money) on these things.

We offer this collection of transactional things at a time when more and more people are interested in taking money apart and putting its components back together in new configurations. The systems of payment that undergird our world are speaking and doing new things—finding affordances in old infrastructures as well as authorizing value transfer as they forge and sunder connections with one another. There is high tech alongside the clunky and even ancient, "disruptive" innovators riding the rails of the legacy systems, a complicated tangle of wires and wishes. It is a good time to explore infrastructures of payment.

Our inquiry into payment comes about also because so many human and technological actors are currently preoccupied not just with exchanges themselves but also with the transactional record of those exchanges. Money itself, as alternative monetary theorists have long recognized, is an archive: a great ledger recording debts and credits. Transactional record keeping is central to money, and so the things of record keeping find a place in our collection, too. As Jane I. Guyer writes in chapter 15, the receipt is "proof of an event in the past"—a proof necessary for the forging of future relations—much like the espresso maker accessed by way of an Airbnb rental (described by Maria Bezaitis in chapter 12) promises a coming together over a steaming cup.

The anthropologist Keith Hart calls money a "memory bank," and the analogy to computing is intentional. For archaeologists of ancient Mesopotamia like Michael Hudson, it goes without saying that before there was coin or paper, there was "book money" (even if the books were made of fired clay). Everyone knows that the paper dollar bill has no value in itself. When they focus on the fiction of fiat currency, they forget that the dollar bill has a serial number; it is a coupon pointing back to an entry in a great, national ledger.

Money is an archive we allow to become a medium of exchange, a marker and store of value referenced back to that archive. Even the most gilded goldbug (or silverites, as explained by Finn Brunton in chapter 20) unwittingly participates in an act of bookkeeping when hoarding gold or Bitcoin against state or fiat money: they are betting that someone, someday in the future, will accept their token in payment for any debt. This is a belief in credit in all senses of the term. Money—any money—needs instruments for noting (or knotting, in the case of khipu, as described by Gary Urton in chapter 7) new entries into the transactional archive. Payment systems provide those instruments.

But transactional archives are growing, becoming a source of value unto themselves. They are a species of "big data," but potentially the biggest data of them all: the complete record of all human and nonhuman transactions involving value. We think it is a good time to look at the other techniques that humans have created to record and figure transactions, from France's Minitel, discussed by Julien Mailland in chapter 14 to the soon-to-be-forgotten magnetic stripe, whose story David L. Stearns tells in chapter 4.

Opening up payment opens up questions of the unit of account, of how accounts are reconciled, of the authority authorizing account reconciliation. Is it the state? A private entity? The people or a community? An algorithm?

Our project seeks to authorize another kind of accounts keeping by curating and convening, putting side by side the object ontologies of this and that system of accounts keeping and settling, from tattooed arms to the sharing economy. We believe it is especially relevant at a time when many seek to create alternative authorities to the unit of account, from Occupy to Bitcoin, American Express to Airbnb.

The authors we have assembled are diverse, and their work has had unexpected influence. David Graeber is an anthropologist whose historical study of debt helped galvanize the global Occupy movement. Bezaitis is a comparative literature scholar whose career has been spent in the information technology industry. Among other things, she has translated Jane Bennett's vibrant materialism into Intel Labs. Urton, a scholar of pre-Colombian archaeology, has influenced computer scientists fascinated with ancient and alternative coding systems. Our scope is audacious: all of human history. We juxtapose archaeological, historical, and ethnographic material. It is a new wonder cabinet of transactional things.

The Chapters in This Book

The book grew out of an experiment: we invited a carefully selected group of individuals for two days of discussion on transactional things. In addition to archaeologists, anthropologists, economists, computer scientists, and science and technology studies scholars, the group included Federal Reserve payment experts, industry representatives from payment start-ups, those who had built the 1970s' interbank networks that became Visa and MasterCard, information technology professionals, and others. After the workshop, we invited participants to write a short piece about their "favorite" transactional thing. We then invited others to contribute to the project as well, resulting in a compendium of payment objects and transactional things.

Transactional objects are visible (cash railways with transparent chambers for shuttling money and receipts across a business establishment; tattoos on the arm of a California chief for measuring standard lengths of strung shells). Their visibility supplies confidence. But transactional objects are also secret or hidden. Proprietary networks carry valuable and sensitive information: what we see on the surface (a Visa card) hides an acquiring and issuing bank, payment processor, or point-of-sale manufacturer. Ben Turner, the barista and money performance artist discussed by Alexandra Lippman in chapter 9, named his "occasional coffee shop" Squamuglia after the secret mail service in *The Crying of Lot 49*. Cash and coin bear material traces of human handling, which carry secret stories of their past trajectories. It is for this reason that Turner says washed money is "less vibrant."

Is "vibrancy" the primary innovation of private currency Dogecoin? As Sarah Jeong describes in chapter 6, it is a cryptographic currency that emerged from a popular Internet meme that depicts the inner monologue of a "cheerfully wary" Shiba Inu dog. Because it mocks and affirms Bitcoin along with the mix of radical libertarian and anarchist politics associated with it, Jeong writes that Dogecoin became, for a moment anyway, the "Internet's official currency."

Are the material traces of interaction lost with electronic forms of payment? We think not; it just takes a little excavation to discover them. We are inspired by archaeology and ontography as well as how ancient practices impinge on contemporary payment ecologies (from tribute to corvée labor). There is also a biographical quality to the essays that follow, such as Mailland's teenage memories of text-based porn accessed via the French state-run Minitel, or Guyer's still-remembered grocery cooperative member number.

Personal histories intertwine with payment technologies, and sometimes the latter are literally inscribed on the former, as in chapter 3 in Lynn H. Gamble's essay on the measurement of wealth among Native Americans in what became the state of California. Payment objects are intimate objects, carried close to the person—or even inscribed onto the person—and in an age of big data, carrying a wealth of data about the person, constituting the person as data. As Keith Hart's contribution shows in chapter 17, payment creates intimacies, individual or group, short or long term, kin and criminal. In chapter 5, Taylor C. Nelms's discussion of group and individual receipts in Ecuador reveals that the receipt serves as an interchange between the informal and formal, all kept safe and secure in a cash box. Payment entails vulnerability after all: during the tense waiting for authorization, for clearance and settlement to take place, as well as in the moments of relationality forged through the gift and the expectation of its return.

A number of our contributors demonstrate the legacies of infrastructural history, how the tracks laid down and paths not taken led to the endurance, even in today's world of virtual touch screen keyboards, of things like the telephone keypad (Michael Palm, chapter 18) and PIN number (Bernardo Bátiz-Lazo, chapter 16) as the signature of payment. Other elements of contemporary payment, such as the thirty–seventy revenue split between application developers and Apple, derive from Minitel's French (pre)modern, its exacting of tribute back to the center, just like Jean-Baptiste Colbert's network of roads—since if all roads lead to the center, hierarchy endures (Mailland).

Questions of liability, risk, and distributional justice are essential not just to money but also the rails that money rides on. In chapter 8, Swartz probes at the racial reality of midcentury dreams of a cashless society. If government benefit payments no longer come by way of a check but rather a prepaid Master-Card (Lisa Servon, chapter 2), then the labor and expenses associated with those payments are off-loaded onto the citizen. One might be quick to reference here the so-called sharing economy, discussed by Bezaitis, where the risk of, say, faulty wiring in an apartment or an aging gas line to a stovetop fall on the user—sharer, beware!

And then there are the roads not taken. Stearns's history of the magnetic stripe on the back of plastic payment cards reveals the old struggle between merchants and banks, with the latter winning the payment wars—for a time—by instituting their preferred technology over the merchants' optical scanning devices. Massive credit and debit card breaches in the mid-2010s would seem to portend the extension of chip-embedded payment cards to the last giant holdout against that technology, the United States. At present, the United States has adopted chip and signature rather than chip and PIN, which is the standard elsewhere. Will US cardholders continue to sign their names (or as Maurer describes in chapter 10, artful scribbles) to authorize transactions? Or will they, like the rest of the world, someday enter PINs? Chip and PIN, of course, brings us back to Palm's essay and the telephone keypad: we are now at an historical moment where a growing number of humans in the developed world will never have experienced pushing buttons on a landline phone and may marvel at the peculiar layout of the keys.

Throughout the history of payment, fraud and fakes have always been a concern. In chapter 13, Whitney Anne Trettien looks at money in the early American republic. Benjamin Franklin employed leaf prints on notes as an anticounterfeiting device. As she notes, the leaves of money captured the tension between organic life and the inorganic reproduction represented by money's own increase—through trade and interest payments. Similarly, as Rachel O'Dwyer demonstrates through her exploration of ether in chapter 19, there is an ever-present dialectic between reality and abstraction in monetary affairs, substances, and networks, visible and invisible. Nature printing recalls the other stuffs of nature in transactional worlds, such as the ether-like shells and tattooed wrists in Gamble's essay, or Urton's discussion of khipu, an encoding mechanism and nonscriptural accounting system for managing the relations of tribute and empire.

Trettien writes that plants' "incessant *becoming-something-else*" made it impossible for a forger to recapture the lost moment in time when a leaf was first pressed on a metal printing block. This, it seems to us, also describes the things of payment, their technologies and ephemera. As records of the transactions of ongoing social life, they only momentarily freeze interaction as it wends its way to whatever comes next.

This collection, then, is also an effort momentarily to freeze the worlds of transactional things so that we might marvel at them, reflect on their transition and effacement, their innovation and old habits, while new rails are being imagined and actualized alongside the old. Payments and their things are transient—the transaction is, after all, "settled"; it ends. Until the next one begins.

This book is not a comprehensive crash course on payments. Instead, it is an invitation to pay more attention to the ecology of transactional things that ultimately make economies possible. It is also an invitation to think about those artifacts and economies differently.

An exhibit catalog serves as a guidebook and memento. Take this collection as an invitation to explore transactional things. Take it as a memory bank, too—a series of pages from the great archive of payment.

For Further Reading

Arestis, Philip, and Malcolm C. Sawyer, eds. *A Handbook of Alternative Monetary Economics*. Cheltenham, UK: Edward Elgar Publishers, 2007.

Benson, Carol, and Scott Loftesness. *Payments Systems in the U.S.: A Guide for the Payments Professional*. Menlo Park, CA: Glenbrook Partners, 2010.

Boellstorff, Tom, and Bill Maurer. *Data, Now Bigger and Better!* Chicago: Prickly Paradigm Press, 2015.

Gibson-Graham, J. K. *The End of Capitalism (As We Knew It): A Feminist Critique of Political Economy*. Minneapolis: University of Minnesota Press, 2006.

Graeber, David. *Debt: The First 5,000 Years*. New York: Melville House, 2012.

Hart, Keith. *The Memory Bank: Money in an Unequal World*. London: Profile Books, 2000.

Hudson, Michael. Introduction: The Role of Accounting in Civilization's Economic Takeoff. In *Creating Economic Order: Record-Keeping Standardization, and the Development of Accounting in the Ancient Near East*, ed. Michael Hudson and Cornelia Wunsch, 1–22. Bethesda, MD: Capital Decisions Ltd, 2000.

Zelizer, Viviana. *The Social Meaning of Money: Pin Money, Poor Relief, and Other Currencies*. New York: Basic Books, 1994.

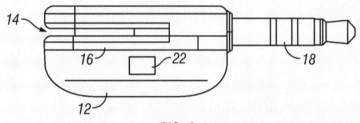

FIG. 2

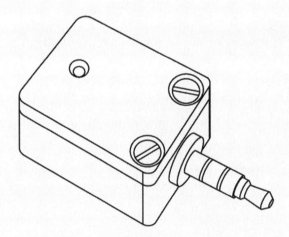

FIG. 3A

1

DONGLES

SCOTT MAINWARING

1.1 Square dongle

I recently ducked into a small independent coffee shop for a quick midday recaffeination, placing my order with the barista/cashier staffing the shop. She took my order, I handed her my IHG Visa rewards card for payment, and she swiped it through a Square Reader plugged into an iPhone. And she swiped it again, reversing direction. She rubbed the card's magnetic stripe on her shirtsleeve and tried again—and again. Muttering something about finicky technology, she unplugged the reader and started rummaging through a drawer. She found another reader, plugged it in, and swiped again, and several more times. After more muttering, she dove back into the drawer, and after further rummaging retrieved a third reader and plugged it in. The transaction went through on the first swipe. "You found the magic dongle!" I said, and she laughed, handed me the iPhone so I could sign an X with my finger, and went off to pour my coffee.

Had I been in a hurry, or had there been a line of people waiting behind me, this extended interaction might have been pretty annoying. As it happened, it was just kind of amusingly quirky, appropriate to the aesthetic atmosphere of the little café. Square, for better or worse, affords this kind of quirkiness, in large part due to its reliance on its reader dongle. Dongles like Square and its competitors underlie a particular genre of technology-mediated payment experiences. Like any genre, donglified payment takes on its character through social and technical features that structure its prototypical examples, and differentiate it from competing genres. Square and its ilk are worth considering for their own sake, as important and potentially long-lived market trends, but also as pointing to larger questions: What is a

1.2 Image macro drawn from Best Buy advertisement, 2015

"payments genre"? How could these be understood by applying multiple perspectives?

The origins of the term "dongle" are unclear. It conjures up "dangle" (and alas, "dong"; see Garber 2013 and Sklar 2013), but the earliest dongles circa 1980 didn't dangle; they were circuit boards or other hardware units that plugged into a personal computer or workstation to authorize it to run associated copy-protected software. In those days long before the Universal Serial Bus (introduced in 1997), these hardware modules would plug into ports intended for cassette tape drives or printers. In this sense, they were "kludges" or hacks: work-arounds that solved a problem in a clever but inelegant or unintended way, often hijacking an existing connection on the host device to use it for its own unorthodox purposes.

From their earliest instances, then, dongles were a way to give special powers to a computer, such as allowing it to unlock protected software, but with a particular aesthetics suggested by the word "dangle." Dangling is not elegant. Rather, it tends to look silly and even precarious—perhaps evidence that the dangler has not thought things through. Calling something a "dongle" can be a way of apologizing in advance for its potential shortcomings, thereby setting expectations appropriately low. If it breaks, you should be prepared to get another. It might not attach securely and so will be left behind; it is your responsibility to keep track of it. It is telling that manufacturers of trusted dongles, like Square, avoid the word "dongle," even if their customers and critics commonly apply it.

By the time the Square Reader debuted in 2011, "dongle" was no longer strongly associated with authorization but instead with connection and adaptation. Need to connect a Macintosh to a conference room projector? You had better have remembered your dongle. Need to connect your laptop to a mobile phone network in the absence of Wi-Fi? Your wireless carrier will sell you a dongle for that. Need to connect your phone to the credit card network? Square's card reader dongle fits right into these schemes. Dongles in these contexts seem less about granting devices special powers and more about connecting two worlds together: Apple devices to PC-centric infrastructure, laptop computers to cell phone infrastructure, and smartphones and tablets to card payment infrastructures. They are the digital equivalents of the electric socket adapters that international travelers must remember to bring with them to be able to make their incompatible devices compatible. Inelegant but functional, they call attention to incompatibilities and seams between materials; in fact, they are the seams.

The Square dongle is no ordinary connector. It is a remarkably clever hack. For many years, payment industry watchers had been waiting for Near-field communication capabilities to be built into smartphones and tablets, ushering in a promised land of tap-to-pay interactions. Square wisely decided not to wait. It created a peripheral that its smartphone app could use to read standard debit and credit cards. This peripheral could have taken a form along the lines of a docking station, communicating with the mobile host through "proper channels" via its docking connector. But docking connectors are notoriously device specific, across product families (Apple versus Android) and even within a product family (e.g., iPhone 4S as well as earlier versus iPhone 5 and later), necessitating providing and supporting different hardware for different users.

Instead Square invented a dongle that could work across all these devices, relying only on a standard 3.5-millimeter, three-contact headset/headphone plug and way from within the device's operating system to access it. Having evolved out of music players and phones, all these devices took for granted that audio would need to flow into and out of them, and in a rare consensus around a single standard had settled on a standard headset/headphone connector. The Square Reader would merely need to listen to a card being swiped as if it were audio being played, and pass this along through the standard audio interface. It turned out that the reader could use inexpensive, off-the-shelf cassette tape read heads for this purpose (see Long 2012a, 2012b). As it turned out, the original dongle provided a bit too direct a connection to the magnetic tape data, facilitating credit card skimming; subsequent versions added encryption and used specialized read heads (see Vanhemert 2013).

Thus the revised Square Reader represents all three facets of dongleness covered so far: it's a hack, cleverly repurposing the audio port of mobile devices for its own ends; it's a kind of authentication token, without which the Square Reader's app will not function; and it's a connector, bringing together two technical worlds.

There are other aspects of the Square dongle (and dongle-based mobile apps more generally) that could be discussed—many having to do with business implications as well as competitive powers and vulnerability. As much of the media coverage of Square takes place from this perspective, I won't comment here beyond pointing out that the business interests of the dongle maker and host maker do not necessarily align, and the role of the patented dongle itself in the business viability of the larger enterprise may be that of a relatively minor, if necessary, piece (for example, as of this writing Square is in a difficult competition against larger, more established players like PayPal and Intuit, each with their own card reader dongle).

To return to the question of aesthetics and payment genres, technical and business considerations are important, but even more crucial are the experiential and social aspects of dongle-based payments that they enable and shape. Here, it is useful to see the Square Reader as not only embedded within a history of the dongle but also within a history of personal device (and particularly mobile phone) decoration and ornamentation.

A brief foray into Amazon's product categorization tree under "Cell Phones & Accessories" uncovers a category called "Phone Charms" with over two hundred thousand entries, next to over twelve million "Basic Cases" and eight hundred thousand "Wallet Cases." Mobile phones are intensely personal devices often used in public settings, so it is hardly surprising that a thriving fashion industry of cases and ornaments has grown up around them. The Square Reader cannot help but be seen in this context, and not just because there are products like "Pouch for Square®" (a key-ringed, faux leather case for the reader available in white, black, pink, purple, and red). Just connecting it to a smartphone or tablet is, among other things, an act of fashion. Unlike the point-of-sale terminals with which it (and other payment dongles) competes, the Square Reader is not, and does not look like, a business technology. Many have remarked on how the industrial design fits into the iPhone and iPad aesthetic of simplicity and rounded corners. The way it ornaments its host device is understated. The device with a reader attached becomes, functionally, a point-of-sale terminal, but it does so without negating the consumerness of the technologies comprising it. The same is even true of the Stand Square, the $300 swiveling iPad holder with built-in card reader, designed for countertop use in small businesses (see Wohlsen 2013).

Transacting with a seller using either the Square Reader or Square Stand feels disarming, especially compared to transactions mediated with conventional point-of-sale terminals and cash registers. It is disarming because it attempts to place the buyer and seller at more or less the same level, saying, in effect, "Here's my smartphone (or tablet). Let's share it to make something together." It's also disarming in asking the buyer to "sign" with their finger on a touch screen, implicitly acknowledging that you're not being asked for your "real" signature, just a gesture in that direction and something that feels almost childlike. There is a kind of enforced informality and playfulness to the whole affair not often present when traditional technologies of commerce are involved.

The Square-Reader-mediated transaction is an impressive achievement, at least in the fairly upscale small or microbusiness contexts in which the Square is found, successfully creating a new genre of payment with its own distinctive feel. It draws on the connotations of dongles as authorization tokens, connection technologies, and clever hacks. It sets up a space in-between high informality (as when friends or family members create and settle small debts among themselves) and efficient anonymity (in which the transaction has as much human feeling as a vending machine). It captures the "sharing economy" zeitgeist in which commerce is reimagined as a conversation between peers rather than a transaction (exploitative or at least impersonal) between subject and corporation. Extending beyond payment per se, many visions for the future of retail are similar imaginings in which buyers and sellers interact as if friends, on the sales floor or in the bank lobby, frequently making use of shareable devices like tablets to flatten differences in social rank.

As with any genre, however, there are limitations, assumptions, and distortions. For one thing, the Square vision (and many visions coming out of Silicon Valley) assumes a world in which class differences don't exist or are hidden, in which everyone has a smartphone and credit card (or at least a debit card), in which disarming interactions in which the cashier hands you the cash register are comfortable not suspicious, and in which the buyer is justified in trusting the seller (and vice versa) this way. This is a bourgeois vision and well aligned with many of the small businesses in Square's market. It also need not be destiny, as this genre depends on two people interacting face-to-face; in that respect, they may be "equals," but the people involved can jointly decide that one party is subservient to the other, or perhaps to pretend for the purposes of the transaction that they are peers when each knows the larger social reality is much different. There are limits to these illusions of equality, however. Square is not well designed, say, for situations in which cashiers are behind bars and bulletproof glass.

Class differences aside, Square transactions also continue to hide power imbalances in the underlying payment rails on which they ride. At the end of the day, a Square transaction is a debit or credit card transaction, and is as fair or unfair as the interchange fees incurred. If it has created a distinctive genre, it is a subgenre of this established and entrenched domain in which transaction fees are hidden, as are the consequences for the consumer of taking on debt. This is only to say, perhaps, the obvious: that Square isn't by itself a disruptive or revolutionary payment experience, in the way that Pay by Square might be or Bitcoin/Blockchain enthusiasts imagine. Nevertheless, it is different enough from what came before to be called, I think, a new genre, and one that could inspire others.

References

Garber, Megan. 2013. The Origin of the Word "Dongle": 7 Leading Theories. *Atlantic*, July 29. http://www.theatlantic.com/technology/archive/2013/07/the-origin-of-the-word-dongle-7-leading-theories/278180.

Long, Evan. 2012a. iPod Meets Reel. http://www.evanlong.info/projects/reeltoreel.

Long, Evan. 2012b. iPod Meets Reel. https://www.youtube.com/watch?v=9QmCfwyA2wc.

Sklar, Rachel. 2013. The Firing of Adria Richards Looks Like Kneejerk Appeasement to the Troll Armies. *Business Insider*, March 22. http:/www.businessinsider.com/rachel-sklar-on-adria-richards-and-sendgrid-2013-3.

Vanhemert, Kyle. 2013. How Apple's Lightning-Plug Guru Reinvented Square's Card Reader. *Wired*, December 9. http://www.wiredcom/2013/12/the-new-square-reader-a-look-at-how-gadget-guts-are-designed.

Wohlsen, Marcus. 2013. New Cash Registers Are Sexy, but What's Beneath the Counter Matters More. *Wired*, May 15. http://www.wired.com/2013/05/square-cash-register-is-cool-infrastructure-is-cooler.

Treasury of the United States

2

CHECKS

LISA SERVON

2.1 Ida May Fuller receiving the first Social Security check on January 31, 1940

In 1939, amendments to the Social Security Act added two new categories of benefits: payments to the spouse and minor children of a retired worker (so-called dependents benefits); and survivors' benefits paid to the family in the event of the premature death of a covered worker. This change transformed Social Security from a retirement program for workers into a family-based economic security program.

On the last day of January 1940, the first monthly retirement check was issued to Ida May Fuller of Ludlow, Vermont, in the amount of $22.54. Miss Fuller, a legal secretary, retired in November 1939. She started collecting benefits at the age of sixty-five and lived to be one hundred years old, dying in 1975. Fuller worked for only three years under the Social Security program, paying a total of $24.75 in accumulated taxes. Her initial monthly check of $22.54 nearly covered her entire contribution. During her lifetime she collected a total of $22,888.92 in Social Security benefits.

Although Social Security is a universal program in that it is available to anyone regardless of socioeconomic status, it does tie benefits to work, reinforcing the notion that a person's value to society stems from their economic productivity. Social Security, which is pegged to inflation, was allocated a $773 billion budget in 2012. It is much more highly funded than means-tested programs, such as "Food Stamps" (Supplemental Nutrition Assistance Program or SNAP) or "welfare" (Temporary Assistance for Needy Families or TANF), which are not tied to inflation. This differentiation between more highly funded universal/worker programs and less highly funded, targeted/nonworker programs reinforces the distinction between the deserving and undeserving poor.

The Social Security Act was amended in 1956 to provide benefits to disabled workers age fifty to sixty-four and disabled adult children. Four years later, in September 1960, President Dwight David Eisenhower signed a law amending the disability rules to permit the payment of benefits to disabled workers of any age and their dependents.

By the late 1940s, machines at the US Department of the Treasury's disbursing centers could print up to seven thousand checks per hour. A decade later, the apparatus necessary to create and distribute Social Security checks had grown to include an array of machines, pictured below. The federal government anthropomorphized these machines in its ads and gave them nicknames. From the top left, they were known as "Punch," the "Interpreter," "the Hen," "Tabby," "the Sorter," and "the Eye."

2.2 1955 Social Security Administration booklet

The check itself became more elaborate, too, as the Treasury added features to cut down on fraud, evolving to the current, multicolored version that shifts from green to yellow to orange. The Statue of Liberty stands guard on the left side of the check, with her torch framing the Treasury's name and seal.

2.3 Social Security check

The Social Security check may soon become a payment artifact, not because the program is ending, but because of a shift from paper to electronic payments. In order to reduce costs, the Social Security Administration began retiring paper checks in May 2011, replacing them with a direct deposit service and the "Direct Express" MasterCard debit card. Testifying before Congress, an administration official stated, "They [electronic payments] are inexpensive—it costs the government about $1.25 to issue a paper check; conversely, it costs only about $0.09 to pay a federal benefit electronically." The switch therefore creates a significant savings for taxpayers. Social Security Administration officials also argue that checks are more susceptible to theft and fraud than are electronic payments.

By January 2013, 93 percent of recipients were receiving their benefits electronically, but the remaining 7 percent required the generation of five million checks at a monthly cost of $4.6 million. By March 1, 2013, all Americans receiving Social Security, veterans' benefits, and Supplemental Security Income were to have signed up for electronic payments.

But not all seniors are interested in the new forms of payment. Many do not have bank accounts. Accustomed to paper checks, they like the familiarity of the object in itself. Exceptions are being made for those born before 1921, and in "rare cases," those living in remote areas without access to sufficient banking infrastructure and those for whom electronic payments would impose a "hardship due to mental impairment." For everyone else, the plastic Go Direct card has already come in the mail.

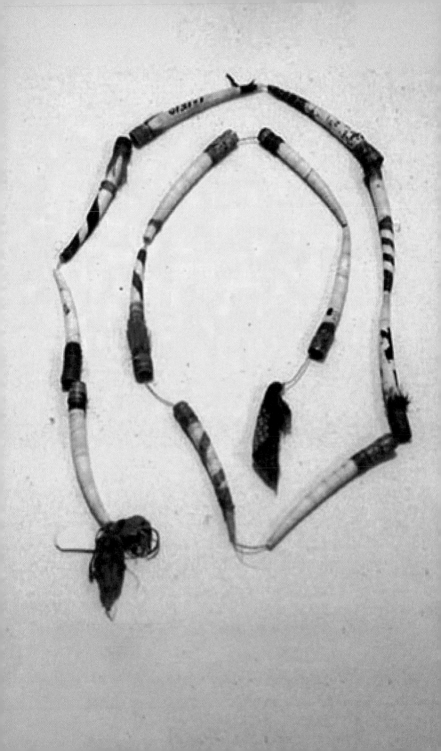

3

TATTOOS

LYNN H. GAMBLE

3.1 Hupa string of dentalia

One of the earliest types of money in the world is the shell bead. In what is now California and elsewhere, bead makers had tedious jobs that entailed gathering shells, breaking them open, and chipping them into rough disks. Once rough disks were roughly formed, they were then perforated with stone drills, strung, and finely ground down. It could take hours to make a handful of beads, some so tiny that they easily could slip through a window screen. California Indians produced hundreds of thousands of beads from the shells of more than twenty-two species of sea animals, including abalone, clam, purple olive (of the genus *Olivella*), and tooth shell (genus *Dentalium*). Dentalium shells differ considerably from disk beads because they usually were not altered; they naturally resemble miniature elephant tusks, and are between 1.5 and 3 inches in length. None of them required drilling. Some were incised and decorated (figure 3.1), but their value was primarily based on their length, not their decoration. Large dentalia are rare and worth much more than smaller shells of the same species. Many dentalia found in California probably originated off the coast of Vancouver Island, over five hundred miles away.

Measuring the value of dentalium shells was critical—just as it is for coinage and other forms of money today—and based on the length of each shell. Among the Yurok Indians in northwest California, a single dentalium shell that was 2.5 inches long was worth about US$5 in the early twentieth century (approximately US$100 today). They were strung eleven to a string. California Indian men of nobility, such as the northern California Hupa Indian named Mr. McCann (pictured in figure 3.2), often had a series of dark lines

3.2 Mr. McCann measuring dentalium shell money against tattoo marks on his forearm

tattooed on the interior of their forearms. Each of these marks denoted the length of a string of dentalium shells (Goddard 1903-1904) that was measured on the arm by holding a string between the thumb and one of the tattoo marks. Slightly smaller dentalium shells were also strung with only eleven per string, but the string was shorter and measured to a different tattoo mark. There were ten such lengths of dentalia strings, and therefore ten tattoo marks on the arms of Hupa men. The Tolowa Indians, neighbors of the Hupa, used a similar system of measurements using tattoos.

The Northwest Coast Indians were not the only ones who used tattoos to measure lengths of strung beads, however. The Cahuilla Indians, who lived near the modern town of Palm Springs, in the south of California, also marked their arms (Strong 1929, 94–96, 107). The beads they traded for, however, were made from the olive shells (*Olivella biplicata*) (King 1995) that were made into disk beads. Produced predominantly by the Chumash Indians along the Santa Barbara Channel coast, these beads were one of the most common types in California (Arnold and Munns 1994; Gamble 2011; King 1976). Their diameters, hole size, and morphology changed over time, yet beads from within the same time period were highly standardized—so much so that we can date most California shell beads by measuring these characteristics, and we know the Chumash made them because of this high degree of standardization. Ancient Chumash villages are filled with remnants of bead-making activities. Partially drilled beads, small stone drills, and thousands of broken olive shells litter the surface of many archaeological sites on the California Channel Islands of Santa Cruz, Santa Rosa, and San Miguel, off the coast of Santa Barbara, attesting to the significance of bead making on these islands. No other region in California has as much evidence of bead making as that seen on these three Channel Islands. Remains of bead making are so profuse on Santa Cruz Island that it was known as the "original mint." Once produced, Olivella beads were widely traded, showing up in sites throughout California and beyond, including in the Great Basin, Colorado

Plateau, Columbian Plateau, and Southwest (King 1990). This brings us back to the Cahuilla Indians, further south near what is now Palm Springs, who consumed many such beads in their ceremonial network.

Unlike the Hupa in the Northwest, where most men of importance had tattoos for measuring beads, the *only* Cahuilla who tattooed their forearms for this purpose were the clan leaders—easily recognized by their distinctive tattoos (Strong 1929, 94–96, 107). Compared with the Northwest Coast groups, the Cahuilla Indians were considerably more restricted in their use of shell beads; shell bead money was associated with ceremonies and owned by the clan, not individuals. The clan leader (*net*) had the great responsibility of keeping the shell beads, along with the sacred bundle, a woven mat containing ceremonial objects. Trade was not individualistic but rather between clan leaders; a clan leader sent strings of shell beads to another clan leader on the death of a clan member (ibid.). The length of the shell beads was determined by wrapping the string of beads around the left wrist, thumbs, and fingers, ending at a location on the wrist four inches from the base of the palm of the hand. This spot, called *tcic'hiiunut*, was marked by a tattoo that was given to the clan leader during a ceremony headed by a ritual specialist. A cactus thorn was used to insert charcoal and make the mark permanent—in this case, just one line. Each clan leader returned the long strings of shell money when another neighboring leader gave a ceremony, thus serving to keep the shell beads in a perpetual mode of exchange. This is similar to the well-documented Kula ring in the western Pacific noted almost one hundred years ago by Bronislaw Malinowski ([1922] 1962), where shell valuables were traded and eventually gained histories as they moved in a continual circle of exchange.

In the Northwest, valuables were coveted and most well-to-do people owned dentalia. In fact, the pursuit of wealth was such a crucial pastime for the Northwest Coast Indians that they safely kept their beads stored in elk horn purses (figure 3.3). Some Northwest groups firmly believed that persistently thinking about shell money would result in its appearance (Kroeber 1953). For instance, a man might reflect about money until he saw money, sometimes in the form of a giant dentalium, the Great Dentalium (*Pelin-tsiek*), peering at him through the doorway of a sweat lodge—a sign that he would become wealthy. Both men and women needed wealth for many reasons, including to purchase goods and services. If a man wanted to marry, he had to pay a set price for this privilege. A bride from a wealthy family might cost multiple strings of dentalium shells, along with additional treasures, such as boats and rare albino deerskins, all

3.3 Yurok elk horn money purse

amounting to US$300 or more in the early 1900s. Women also desired and owned dentalia, often earning it through doctoring, an important profession for women in this region. Just in one healing session, a woman could bring in the equivalent of US$10–20 in strings of dentalia at the time (ibid.).

In these two California examples, therefore, the use of tattoos to measure lengths of strung shell beads is similar, but the meaning of the beads, the types of beads, their circulation, and the people who possessed them all differed. The concept of wealth questing was foreign to the Cahuilla. On the contrary, in southeastern California, Olivella shell beads were closely tied to ceremonial exchange and not individually owned as they were in northern California. The possession and distribution of shell beads was tightly controlled by clan leaders, and tied to sacred ceremonies. The clan leader physically kept the beads and was in charge of dividing them among invited clan leaders when conducting an image-burning ceremony—an event that was usually held about a year after the death of a prominent person. In contrast, for the Hupa like Mr. McCann and neighboring groups in the Northwest Coast region, the acquisition of valuables was integral to life. There was no dishonor associated with the ownership of wealth. Valuable items such as strings of dentalium shells, large obsidian blades, and dance regalia, including headdresses with hundreds of red woodpecker scalps, were (and still are) proudly displayed during ceremonies and other events. Their treasures were carefully kept inside their houses in large wooden boxes. Indeed, the groups of this region had an extraordinary preoccupation with wealth (Wallace 1978, 169–171).

Shell beads, considered the iconic artifact of what is now California, were produced, traded, and used by Indians in that region for over eleven thousand years. The great antiquity and prevalence of shell beads there, and permanent tattooing of one's body for the purposes of measuring strings of beads, attest to their significance in the economic, political, ceremonial, and social life of the Indian groups of the region. Tattoos on the bodies of men alternately signified social hierarchy or the quest for individual gain—records in flesh of value, albeit different values.

References

Arnold, Jeanne E., and Ann Munns. 1994. Independent or Attached Specialization: The Organization of Shell Bead Production in California. *Journal of Field Archaeology* 21 (4): 473–489.

Gamble, Lynn H. 2011. Structural Transformation and Innovation in Emergent Economies of Southern California. In *Hunter-Gatherer Archaeology as Historical Process*, edited by Kenneth E. Sassaman and Donald H. Holly, 227–247. Tucson: University of Arizona Press.

Goddard, Pliny E. 1903–1904. Life and Culture of the Hupa. *American Archaeology and Ethnology* 1 (1): 1–88.

King, Chester. 1976. Chumash Intervillage Economic Exchange. In *Native Californians: A Theoretical Retrospective*, edited by Lowell J. Bean and Thomas C. Blackburn, 289–318. Ramona, CA: Ballena Press.

King, Chester. 1990. *Evolution of Chumash Society: A Comparative Study of Artifacts Used for Social System Maintenance in the Santa Barbara Channel Region before A.D. 1804*. New York: Garland Publishing.

King, Chester. 1995. Beads and Ornaments from Excavations at Tahquitz Canyon (CA-Riv-45). In *Archaeological, Ethnographic, and Ethnohistoric Investigations at Tahquitz Canyon, Palm Springs, California*, edited by Lowell J. Bean and Sylvia B. Vane, vol. 2. Menlo Park, CA: Cultural Systems Research.

Kroeber, Alfred L. 1953. *Handbook of the Indians of California*. New York: Dover Publications.

Malinowski, Bronislaw. (1922) 1962. Argonauts of the Western Pacific. New York: E. P. Dutton and Co.

Strong, William D. 1929. *Aboriginal Society in Southern California*. Berkeley: University of California Press.

Wallace, William J. 1978. Hupa, Chilula, and Whilkut. In *Handbook of North American Indians, Vol. 8, California*, edited by Robert F. Heizer, 164–179. Washington, DC: Smithsonian Institution.

4

MAG STRIPE

DAVID L. STEARNS

4.1 A wallet with slots designed to hold standard plastic payment cards

I've always been fascinated by plastic payment cards. I've spent many years studying and writing about the systems that lie behind them, but I also love just handling and admiring the cards themselves. I love the way the plastic substrate is light yet remarkably durable, able to take on any kind of graphic design imaginable and still scrape ice off a windshield (albeit poorly). I'm impressed by the subtle rounding of the corners, which not only make the card comfortable to handle but also easy to slip into and out of a wallet or restaurant bill sleeve. I like running my fingers across the embossed card numbers, vestiges of an earlier era of multipart sales slips and card imprinters (which can still be used when electricity or telecommunications are not available). And I love staring into the inky blackness of that much-maligned magnetic stripe across the back, a simple though highly insecure solution for making the card machine-readable. Payment cards are quite complex artifacts if you look at them closely, packed with social meaning as well as technical capabilities.

Although I hate to admit it, my fascination with these cards is most likely a form of proleptic nostalgia. Those of us who study the payments industry know that mobile wallet and payment schemes, which replace the plastic card with a chip and smartphone application, are growing in popularity, especially among the younger generations. Plastic payment cards have been with us for quite a while now (since 1958), but if the popularity of mobile wallet schemes continues to rise, they may soon go the way of 8-track tapes and VHS videocassettes.

In many ways, we shouldn't be surprised by this. After all, the plastic payment card isn't like a banknote or even a personal check. The payment card is merely an access device, a means for identifying the cardholder to the vast electronic financial network that lies behind it. This network separates identity authentication from the financial instruments that actually make money move, and the card really only provides the former function, not the latter. While the plastic card has served this function fairly well for over five decades, there is no particular reason why it must continue to do so. In fact, identity authentication could be accomplished much more securely and effectively by a computerized device with multiple sensors, capable of verifying something you have, something you know, and something you are.

Even though most other developed countries have already moved to using cards with embedded computer chips that verify something you know (a PIN), banks in the United States still issue cards with that much-maligned magnetic stripe across the back. The stripe is what allows various machines to read your card, and it is the primary way in which your card identifies you to the payment system. Unfortunately, most of these cards can also be used in places where verifying something you know (entering a PIN) is not possible, allowing thieves to easily use stolen or skimmed cards.[1] Thankfully, federal laws limit consumer liability, but they raise a few interesting questions: Why do US banks still issue magnetic stripe cards, and why did they adopt such a vulnerable technology in the first place?

To answer these questions, we first need to put all of this into some historical context.[2] Plastic payment cards actually predate ATMs and merchant point-of-sale card readers, so as these devices were being developed, the banks had to make a decision as to how exactly these machines should "read" the necessary information off the card so that transactions could be processed electronically. Various ideas were floated, but by the early 1970s, three primary approaches were being actively pursued: optical character recognition (OCR), magnetic stripes, and Citibank's proprietary "Magic Middle" technology.

OCR was an attractive solution, as it read the existing numbers on the face of the cards and thus wouldn't result in any increase in card manufacturing costs. It was favored by the oil and retail industries, as they were already using OCR to transform their paper sales drafts into electronic records. Supermarkets were also moving toward optically scanning bar codes on their products, and including one on the card would make it easy to read its information. Although OCR terminals were still relatively expensive, they had proved reliable, and their costs were quickly falling.

The airline and banking industries took a different approach—one based on magnetics instead of optics. Magnetics was a well-known technology in the banking industry; banks had been using magnetic ink character recognition for automated check processing since the 1950s. In the late 1960s, IBM developed the magnetic stripe for the airline industry, which was looking for a way to make its payment cards and passenger tickets machine-readable, and several large banks started using it on their cards as well.[3] Magnetic tape was a relatively simple, inexpensive, and flexible solution compared to OCR, as the stripe could be divided into several tracks and encoded at different densities. Although it did increase the manufacturing costs a bit, it also offered the chance to encode additional "invisible" information that could be used to restrict functionality or enhance security.

Unfortunately for the banks, audio playback and recording components had become inexpensive and readily available to consumers by the early 1970s, making it easy for criminals to build "skimmers," simple devices that could quickly read, store, and

a.

b.

4.2 The front and back of a 1974 BankAmericard showing an early form of the magnetic stripe replicate the entire contents of a magnetic stripe. The replicated card could then be used to make numerous purchases before the cardholder or issuing bank even knew what had happened.

Although the American Bankers Association (ABA) understood the security vulnerabilities of the magnetic stripe, it nevertheless adopted it in 1971 as the standard way for making a bank card machine-readable. The ABA was looking forward to an assumed future where all merchants and banks would be connected to one another through telecommunication links, and every transaction would be authorized with a PIN by the card-issuing bank's computer. The magnetic stripe might easily be copied, but the thief would still need the PIN. The bank's computer could also easily detect purchases out of the norm and quickly shut off the card.

Citibank, however, had invested more than $30 million in an alternative that it hoped would dethrone the magnetic stripe and become the standard nationwide. It was also Citibank's own proprietary technology, the adoption of which would result in significant licensing revenue for the bank. The solution was known as the "Magic Middle" because it consisted of a piece of reflective material sandwiched between two layers of plastic. The middle layer could be encoded with punched holes, much like the punched cards used by early computers. The Magic Middle cards were read by shining infrared light from one side (which would pass through the plastic and any holes punched in the middle layer) and sensing what came through on the other side. Citibank claimed that this technique was far more secure than the magnetic stripe, and in order to emphasize its point, sponsored a contest at Caltech to come up with the best ways to defraud the ABA's standard magnetic stripe. Not surprisingly, the ABA reacted to this contest rather negatively, amending its card-encoding requirements to explicitly exclude proprietary technologies. Citibank continued to develop the Magic Middle, but found few banks willing to license it. In 1979, Visa drove the last nail into the Magic Middle's coffin when it required all cards issued with its mark to include a magnetic stripe.

The magnetic stripe was hardly perfect, but in the 1970s, it was probably the best option for making cards machine-readable. As authorization terminals became widely adopted in the early 1980s, banks were able to detect and quickly shut off fraudulent cards, lowering fraud rates to somewhat-manageable levels. Because the telecommunications infrastructure in the United States became so reliable and affordable, the banks and card networks decided to put their security efforts into analyzing purchase patterns and quickly detecting anomalies, rather than shifting toward a point-of-sale technology that would prohibit or deter unauthorized use in the first place.

Other parts of the world took a different approach, largely because their telecommunications were less reliable or more expensive. These countries still use plastic cards, but the cards contain a microchip instead of (or in addition to) a magnetic stripe. Chip cards are much more expensive to manufacture, but the chip is able to verify something the cardholder should know (a PIN) without having to send a message back to the issuing bank. Although these "chip-and-pin" cards are not immune from attack, they have significantly reduced fraud rates in situations where the card can be physically inserted into a merchant terminal.[4] In an online purchase situation, though, these cards are no more secure than magnetic stripe cards, as there is no merchant terminal for authenticating the user. Unfortunately, this is also the largest and fastest-growing area of card fraud.

This largely explains why the United States has lagged behind the rest of the world in replacing magnetic stripe cards with chip cards, but this will soon change. Visa and MasterCard

4.3 TRANZ 330 magnetic stripe point-of-sale terminal by Verifone

are actively incentivizing their banks to issue chip-containing cards as well as their merchants to install the requisite point-of-sale equipment (contact based or contactless). Under the old liability rules, the issuing bank absorbed the losses from fraudulent cards, but after October 1, 2015, merchants who did not adopt chip-reading point-of-sale terminals became liable for any fraud involving a chip-based card (fuel-dispensing merchants were given an additional two years to comply, largely because replacing self-service gas pumps is extremely expensive).[5]

Chip-based cards will certainly sound the death knell of the much-maligned magnetic stripe, but I suspect that they may also hasten the obsolescence of the plastic card itself. Once the payment infrastructure shifts to using chips, there is no particular reason why those chips need to remain embedded within rectangular pieces of plastic. Plastic was a convenient substrate for payment cards requiring embossed numbers and a magnetic stripe, but a flexible computing device could interact with a payment chip in far more interesting ways. Some of these ways are already being demonstrated in mobile payment platforms such as LevelUp, which combines payments processing with loyalty and rewards programs, gift cards, preordering, and rich data analytics for the merchant.

Of course, plastic cards won't disappear overnight. Futurists in the 1960s famously predicted that we would achieve a "cashless society" within a few decades, and that has certainly not yet come to pass.[6] Older generations often strongly resist changes to their payment mechanisms, and merchants are typically loath to invest in new kinds of terminals if they don't have to, so any large-scale change will likely take a few generations to occur. Still, I expect that children born fifty years from now, when they see an old movie showing a plastic payment card being swiped through a magnetic stripe reader, will ask their parents, "What is that?"

Notes

1. For a fascinating look into the world of card skimming and fraud, see Kevin Poulsen, *Kingpin: How One Hacker Took over the Billion-Dollar Cybercrime Underground* (New York: Crown, 2011).

2. For a complete history of how the banking industry and payment card networks adopted the magnetic stripe as well as details about the alternatives at the time, see David L. Stearns, *Electronic Value Exchange: Origins of the VISA Electronic Payment System* (London: Springer, 2011), chapter 7.

3. For an insider history of the magnetic stripe, see Jerome Svigals, "The Long Life and Imminent Death of the Mag-Stripe Card," *IEEE Spectrum*, May 30, 2012, http://spectrum.ieee.org/computing/hardware/the-long-life-and -imminent-death-of-the-magstripe-card (accessed September 2, 2013).

4. See, for example, Mike Bond, Omar Choudary, Steven J. Murdoch, Sergei Skorobogatov, and Ross Anderson, "Chip and Skim: Cloning EMV Cards with the Pre-Play Attack," University of Cambridge Computer Laboratory working paper, http://www.cl.cam.ac.uk/~rja14/Papers/unattack.pdf (accessed September 2, 2013).

5. See "Visa Announces Plans to Accelerate Chip Migration and Adoption of Mobile Payments," August 9, 2011, http://usa.visa.com/about-visa/ index.jsp(accessed September 2, 2013).

6. See Bernardo Bátiz-Lazo, Thomas Haigh, and David L. Stearns, "How the Future Shaped the Past: The Case of the Cashless Society," *Enterprise and Society* 15, no. 1 (March 2014), http://journals.cambridge.org/article_ S1467222700000057 (accessed March 16, 2015).

5.1 The treasurer of the Álvarez family caja reviews accounting documents with the author in Quito, Ecuador

5

ACCOUNTS

TAYLOR C. NELMS

Cash is not generally thought of as a cutting-edge payment technology. Cash, it is often said, is dirty; cash is risky; cash is anonymous; cash is dangerous; cash mediates everyday, informal, popular, street, or subsistence economies; cash is for laundering; cash is for storing away in safes and coffee cans. Cash, like the livelihoods it moves through, exists apart. Indeed, despite the fact that many people all over the world (including many in the United States) live cash-only or cash-heavy economic lives, predictions of—and calls for—the elimination of cash have become commonplace today. Digital and mobile payments are thus frequently seen as alternatives to cash. And yet cash remains an integral part of monetary ecologies around the world, and so it is important to understand the diverse but often-mundane ways people use bills and coins.

This includes understanding how cash use intersects with other forms of money and monetary practices. When we think about cash, we do not always pay attention to how people use it in creative combinations with other modes of payment, exchange, reserve, and accounting to create, discharge, and keep track of debts and obligations. But cash as a payment technology is usually surrounded and supported by a variety of other technologies, especially technologies of documentation. Think about the variety of artifacts that accompany cash: the checks, bills, tickets, stubs, files, records, logs, accounts, and receipts that serve in different ways to materialize and preserve chains of cash transactions. The word "receipt" means, of course, the reception of funds, but it has also come to signify the document that verifies or substantiates that reception. It is not only cash that produces such traces; using cash often necessitates their production. For bills and coins are traces only of themselves. They do not point to value elsewhere but instead are themselves the value they embody. Accounting practices, from tally marks to double-entry bookkeeping, have recorded transfers of value as money changes hands since the origins of money—and perhaps, in fact, predate the tangible manifestations of coins altogether. Such documentation serves as external memory device; as confirmation and sometimes evidence; as testimony to a transaction realized and sum transferred; and as visible display of one's status, credibility, and trustworthiness.

In this chapter, I reflect on cash and its accoutrements. I am inspired by the documentary ephemera of cash payments I came across while conducting fieldwork in Quito, Ecuador, with several neighborhood and family savings and credit associations. Such associations are called *cajas* in Ecuador, and here I focus on the caja of the Álvarez family, formed in 1993 by eight siblings along with their immediate and extended families. When I was conducting fieldwork, there were twenty-six members, including the siblings' spouses, children, and grandchildren as well as the family of a cousin. Members made regular contributions to a common pool, which was then lent back out in parts. The goals of the caja, however, were "social" and financial, the members told me, and each month, one of the households hosted the others to settle accounts and catch up with one another. Before the meetings were officially called to order, members approached the caja manager and treasurer to make their monthly contributions and loan payments, handing over folded dollar bills to the treasurer, who punched numbers into a calculator and made change out of an orange cloth sack she kept next to her on the couch. Each payment was dutifully recorded by hand in the caja's account book, the graph-paper pages of which were filled with past treasurers' careful lettering.

After one of the meetings, Sonia, the caja's secretary, let me examine and photograph the book—a notebook with a floral cover bound with string. In the back, fastened into a yellow manila folder, were a series of pages on which were taped receipts from deposits made to an account, shared by members of the caja, at one of the largest banks in Ecuador. Taped on top of duplicate computer printouts, a cache of smooth and fading papers, the receipts recorded a long history of deposits made by the treasurer, including not only the members' savings but also payments they had made toward the construction of a house outside Quito, which they planned to use for family vacations and reunions. The newest of these receipts were bright and clear, while the oldest were barely legible.

This family caja is just one example of many savings groups, lending circles, cooperative consumption organizations, and other kinds of collective financial associations formed in Quito among family members, neighbors, friends, and coworkers. In the literature on economic development and poverty alleviation, autonomous money-pooling organizations are known as rotating savings and credit associations (ROSCAs), or accumulating savings and credit associations (ASCAs). Both associations are often presented as coping mechanisms for the poor, ladders to enhanced economic development, or opportunities to extend financial services to those without access to formal banking institutions. But those terms belie the remarkable diversity of economic practices and organizational principles of these groups. They also conceal the often vastly different socioeconomic contexts in which they operate. Indeed, for Sonia and her siblings, the caja was not about finding an alternative to the bank, nor simply a creative way to make do with limited resources, but also, perhaps more important, a creative way to bring their family closer together, to overcome economic inequality *within* the family. "The organization is one way that we maintain the bonds between us," Sonia told me. "We are trying to make sure that everyone is on the same level, that we can all take advantage of the [same] possibilities."

While cashlessness is seen as an alternative for many in the United States and Europe, in Ecuador and elsewhere, the cash-based practices of local financial organizations like the Álvarez family caja are themselves increasingly viewed as alternatives—not just as a foundation for development, but as inspiration for imagining other kinds of economies, too. In Ecuador, such associations are included in the so-called popular and solidarity economy (*economía popular y solidaria*, or EPS), which is explicitly framed as an alternative to orthodox economic theory and mainstream finance. The EPS is enshrined in the country's constitution, which was approved in 2008 as part of an effort by the recently elected president, Rafael Correa, to remake Ecuadorian state and society, in part by installing a "post-neoliberal" economic plan and development model. It has since become the object of a variety of legislative reforms, institutional transformations, and regulatory interventions by policy makers and bureaucrats (Nelms 2015). In the eyes of both state and civil society actors, organizations grouped under the EPS umbrella are guided by "social" principles and values: cooperation and community instead of competition; autonomy and self-organization instead of dependence; reciprocity, mutual assistance, and solidarity instead of profit seeking and capital accumulation.

The Álvarez caja, however—like many of the entities taken to constitute this popular and solidarity economy—encompassed a plural collection of practices. Sonia and her relatives told me that the caja provided them access to small-value loans and a way to save, but their primary objective continued to be, as Sonia put it, *"la unión familiar"* (family unity)—that is, "bringing the family together and giving us a tool to reinforce our family bonds." Sonia called the caja the "material component of the family."

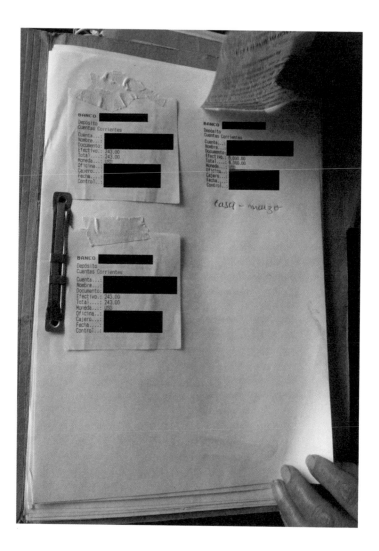

5.2 Deposit receipts taped over photocopies of the same, fastened into a folder held by the treasurer of the Álvarez family caja in Quito, Ecuador

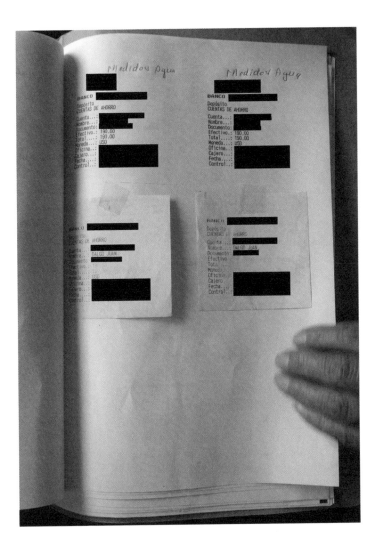

She explained that through the caja, "we have helped our relatives who maybe did not have the same opportunities." The purchase of property and construction of a house outside the city was a part of the "progress" of the family, she told me. Each member of each household contributed financially; all of them benefited from the shared savings and enjoyed the time they spent together. Savings, credit, togetherness, progress, and their future together as kin—the caja served many ends. This multiplicity of ends corresponded with a multiplicity of means: not only accumulating savings and rotating credit, but also a common fund used to reimburse the hosts of the meetings, fund for health emergencies, life insurance fund, and monthly raffle.

All of these financial practices were mediated by cash transactions. But caja members also relied on a bank account. This is not uncommon among such financial organizations; many groups rely on the "formal" banking system to save together "informally." Cash crisscrosses "high" and "low" finance. Often, two members will open a joint checking account; only those with their names attached can make withdrawals. In the Álvarez family caja, the account was opened in the name of the association's treasurer and its president, Sonia's husband. Sonia and her relatives made regular deposits of cash into the account, using it to aggregate their monthly contributions, to facilitate savings for the caja's other funds and make payments toward the construction of the house in the country, which they did via electronic transfer between their account and the contractor's.

At the intersection of these two institutional regimes, the informal family caja and formal banking system, is the deposit receipt. The receipt is a material and visible record of the deposit. It embodied and reinforced the trust that members showed in both one another and the bank where they kept their collective savings. These thin slips of paper, printed to be discarded, were themselves deposited in the back of the account book, layered on top of one another, the hoarded remainders of an unfolding transactional history.

These receipts struck me not only because they are indicative of a more general commitment by caja members to documentation and record keeping but also because the way that they were preserved in the caja's account book suggests a more complex relationship between the institutional regimes of finance—formal and informal, high and low—than typical representations allow. Slips of paper issued by a major bank documenting the receipt of funds saved collectively by family members or neighbors: What better image of the muddled boundaries and crisscrossed scales of everyday financial practice than this? The mainstream financial system does not simply subsume the lives and livelihoods of participants in everyday, informal, popular, street, or subsistence economies; instead, the means of the former—its technologies, channels and pathways for moving money, and legal and institutional infrastructures—are bent toward or folded into the concerns (and accounts) of the latter.

Money is memory. So say the economist Narayana Kocherlakota and anthropologist Keith Hart. For Kocherlakota (1996), money is a record keeping technology, a way to track transactional reallocations of resources. For Hart (2001, 17), money is a communicative medium that serves as a "memory bank" not only for individual transactions but also social interactions, identity, and collective life. It is a "cultural infrastructure" like language itself. Money also leaves memories. For payment circumscribes communities, whether nation or family, and traverses the boundaries within and between them, and as it circulates, money leaves tracks, deposits, and receipts. Even supposedly anonymous cash leaves traces as it travels: credits and debits scratched into account books; receipts spit out by ATMs or handed over by clerks, and saved away until the ink rubs thin and diaphanous.

References

Hart, Keith. 2001. *Money in an Unequal World: Keith Hart and His Memory Bank.* New York: Texere.

Kocherlakota, Narayana R. 1996. Money Is Memory. Federal Reserve Bank of Minneapolis Staff Report 218. http://www.minneapolisfed.org/research/sr/sr218.pdf.

Nelms, Taylor C. 2015. "The Problem of Delimitation": Parataxis, Bureaucracy, and Ecuador's Popular and Solidarity Economy. *Journal of the Royal Anthropological Institute* 21 (2): 106–126.

6.1 The official Dogecoin logo

6

DOGECOIN

SARAH JEONG

Well, a "dollar" is just a meme
of a president—you don't get to
decide what's real!

—Matt Bors, political cartoonist

The Jamaican two-man bobsled team qualified for the Sochi Winter Olympics, but didn't have the money to attend. So they turned to crowdfunding on the Internet. The campaign went viral, and the publicity induced the International Olympic Committee to pay their travel costs. The money from the crowdfunding campaign (eventually over $180,000) helped to offset costs for equipment and training.

In 2014, the success of a crowdfunding campaign is hardly a novel tale. What sets this story apart is the key role played by Dogecoin, a cryptocurrency similar to Bitcoin that plays on the "doge" Internet meme.

Liam Butler, head of the Dogecoin Foundation, was a fan of the film *Cool Runnings*, which is loosely based off the story of how the Jamaican bobsled team qualified for the 1988 Winter Olympics. Motivated by nostalgia for his favorite childhood movie, he started the "Dogesled" campaign, soliciting donations in Dogecoin. The fund ultimately raised about $30,000 worth of Dogecoin, a not-insignificant contribution to the total amount collected.

In July, 2017, a thousand Dogecoin was the equivalent of 27¢ in US dollars, and the total market capitalization of the currency was approximately $25 million. But in 2014, Dogecoin was over seven times more valuable. "In the beginning, everything was awesome and hilarious," said Ben Doernberg, once a board member of the Dogecoin Foundation, in an interview for *Motherboard*.

The story of Dogecoin is the story of Bitcoin in miniature. Each and every cryptocurrency is a referendum on whether money qua money can exist without an issuing government. But each of these grand theoretical experiments always seems to end with some schmuck stealing all the coin and disappearing.

○

Virtual currency or payment systems based on cryptography did not begin with Bitcoin; David Chaum's DigiCash payment system preceded it by over a decade. Yet the Bitcoin protocol is the first implemented decentralized system, eventually spawning multiple derivatives—some serious, like Litecoin and Peercoin, and others more tongue in cheek, like Dogecoin and Coinye West.

6.2 A doge image macro made in honor of the Dogesled campaign

These protocols—called cryptocurrencies—are (usually) decentralized peer-to-peer virtual currencies that track transactions through a public ledger called the Blockchain. The transactions are verified as users' computers broadcast updated versions of the Blockchain to the rest of the network. The Blockchain is finalized one block at a time, with the first computer or group of computers to finalize the block being rewarded with "mined" coins. Miners who expend more computing power, thus doing more "work," are more likely to be rewarded.

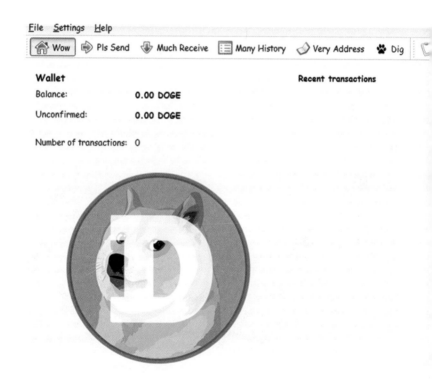

6.3 The Dogecoin client

Bitcoin and its progeny-cryptocurrencies are by design decentralized; the code itself rejects the concept of a "trusted third party," such as a bank, payment processor, or government. The logistics of minting of the coin are preordained by the code, and the work of minting is distributed through the network. When it comes to fraud prevention and transaction disputation, cryptocurrencies simply shift the burden onto the user: all transactions are irreversible. The code does not allow for chargebacks. If a payer is robbed or defrauded, there's no recourse.

The decentralized aspect of cryptocurrencies might be the most confusing one. Cryptocurrencies don't resemble any other currency we know of. No central authority issues the coin, backs it with another commodity, or promises to accept it in lieu of a tax obligation. In a way, the very existence of cryptocurrencies embodies a debate about the nature of money itself: Can money exist separate from an issuing government?

○

Dogecoin's technical deviations from Bitcoin are minor: it is based on a different cryptographic algorithm, and the parameters of coin creation are set differently. For example, Bitcoin is set to max out its supply at twenty-one million coins. The Dogecoin "supply" is set at ninety-nine billion, with an additional five billion to be added every year subsequent to reaching this threshold. But these details are irrelevant to Dogecoin's unusual success in the alternative currency marketplace: the currency's popularity is simply an extension of the doge Internet meme.

The doge meme usually manifests as an image macro of a Shiba Inu dog with colorful text in Comic Sans floating around its head. Although the theme of each image macro may vary wildly, the text almost always adopts a particular grammatical syntax. Typical dogespeak might be "very trubble," "much confuse," and "wow." The overall effect is something akin to disconnected thoughts in the stream of consciousness. Adrian Chen wrote of the doge meme in *Gawker*, "The words are supposed to be the dog's internal monologue, and the dog is astonished, it seems, by the very fact that it is having an internal monologue. (Dogs aren't supposed to have monologues, see.)"

Although many different photographs of Shiba Inu dogs have been used to generate these image macros, the most common one is that of Kabosu, a Shiba Inu belonging to Atsuko Sato, a Japanese schoolteacher and blogger. In this photo, the dog's head is slightly turned so the whites of her eyes are showing. For humans, she appears to be nervously smiling. Her expression has been described as "skeptical," "surprised," "bewildered," "coy," "cheerfully wary," or even one of "shocked delight."

Both dogespeak and the iconic image of Kabosu permeate the branding of Dogecoin. The currency has adopted as its logo a graphic of a golden coin with a Shiba Inu and capital D. The Dogecoin client interface labels menu options with names like "Wow" (home page), "Pls send" (to send to a wallet address), "Much Receive" (to set up one's own wallet addresses), "Many History" (transaction logs), "Very Address" (an address book), and "Dig" (to mine Dogecoins). By fusing Blockchain technology with a meme, Dogecoin straddles the valley between the functional and the inane.

Dogecoin is not the only joke cryptocurrency, but it has been the most prominent one. After launching in December 2013, Dogecoin was characterized by an unusually accelerated rate of adoption, quickly becoming one of the cryptocurrencies with the largest market capitalizations. And despite the fever pitch of media attention on its more serious predecessor, Bitcoin, the number of transactions made in Dogecoin actually exceeded

6.4 A photo of Kabosu, most commonly used in doge image macros

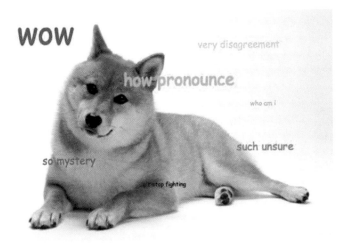

6.5 The pronunciation of "doge" is the subject of much debate; according to a 2013 poll in *Slate*, "dohj" has a slim majority over "dogue."

that of Bitcoin in January 2014. These transactions—when they are between two different people rather than the same person shifting coins between wallets—are mostly quite small.

On Twitter and Reddit, Dogecoin "tipbots" allow users to make microtransactions embedded in the flow of social interactions. Users will "reward" other users for posting what can be described as socially valuable content—anything ranging from funny jokes to more covert interactions, like the distribution of sexually explicit pictures on the Subreddit Girls Gone Doge. Tips can range between ten to five hundred Dogecoin (from below one to about six cents in US dollars).

Dogecoin can be and is converted into real goods and services. The most visible Dogecoin transactions, however, are these miniscule tips, which carry more social value than economic value. In a way, when one responds to a funny comment with a Dogecoin tip, the payment is one in kind: a joke for another joke.

○

In December 2013, software developer Steve Klabnik wrote a blog post on how Dogecoin, as "an absurdist art project, mocking

sarah jeong
@sarahjeong

🐦 Follow

@tipdoge **tip** @lanalana **100 doge**
8:48 AM - 20 Feb 2014

↰   ⇄   ★ 1

6.6 The author tips the editor of this volume a hundred Dogecoin
Source: Screenshot by the author

the entire global financial system," changed his perspective on money itself. He was struck by the ambiguous "realness" of Dogecoin; its jokey character led people to ask, "Is Dogecoin real?" But that question itself evolves into a critical interrogation of whether any money is real, and what realness even is. Dogecoin preempts such a critical interrogation of its socially constructed value by openly making fun of itself. Klabnik commented,

> If things really are worth whatever anyone says they're worth, so are memes. Do we really want to live in a world where a dumb joke about dogs that can't talk has the same legitimacy as money? Because the more people that take Dogecoin seriously, the more serious it becomes, ... and the very nature of its silliness encourages us to think seriously about the nature of money and therefore value.
>
> Before, I just laughed. Now, thanks to a joke, I'm scared.

At the bottom of his post, Klabnik included his Dogecoin wallet address, in case anyone wanted to tip him.

○

The open absurdism of Dogecoin stands in stark contrast to the humorless rhetoric that has surrounded Bitcoin since its early days. As noted by Maurer, Nelms, and Swartz (2013), Bitcoin is often compared to material currencies like gold. Bitcoin's pseudonymous creator, Satoshi Nakamoto, compared Bitcoins to gold in his original white paper, and even used the term "mining" to describe the mechanism through which new Bitcoins are added to the total money supply. Today, Bitcoin proponents still appeal to the idea of "sound money" by pointing out the intrinsic limitations imposed by its parameters. When explaining his and his brother's decision to invest in Bitcoin, Tyler Winklevoss (2013) told the *New York Times*, "We have elected to put our money and faith in a mathematical framework that is free of politics and human error."

Dogecoin, on the other hand, is about as far from practical materiality as is possible. Its value stems in large part from the ineffable social value of humor. It subverts the serious imagery of a gold coin by putting a dog's face on it, and mocks the rhetoric of "mining" by relabeling it as "Dig" in the official client. Even though Dogecoin piggybacks off Bitcoin's design, the rhetoric of "soundness" through a mathematical framework is absent from its narrative. Accounts in mainstream media frequently use the phrase "based on an Internet meme," which gives the uncanny impression of a currency pegged to a viral but essentially frivolous idea.

○

Not even a joke currency is immune to internal monetarist-Keynesian schisms regarding inflation. The Dogecoin parameters have been a source of contentious debate. Where Bitcoin has a strictly limited supply of coins, Dogecoin will be dispersing an additional five billion a year after reaching its supply cap. Heated discussions on both Reddit and GitHub (a code repository) raised complex analyses of the ideal inflationary rate for a cryptocurrency, balancing various factors like the rate at which the currency was being adopted by new users, rate at which coins were being dispersed, changing value of the coins, and rate of inflation anticipated by "real-life" sovereign currency. A user in favor of the "rolling supply" wrote:

When a coin is deflationary, people hoard it, and stop using it. After a while, the circulation of the coin is so low that a single hoarder trying to cash out causes a crash because there is no-one to buy. Initially the coin recovers quickly because other hoarders will see a buying opportunity, but if the confidence in the coin erodes, there is no "real" economy (people buying stuff with the coin) to cushion the blow when the crash comes. A coin that is hoarded is ultimately a zero-sum game: for every dollar someone gains on it, someone else has to lose a dollar when everyone runs to the exit. Only with a sustainable economy where thousands of people have coins in their wallets for daily use can have lasting value.

When explaining the decision to stick with a rolling supply, a Dogecoin developer was careful not to take sides on whether inflation was good or bad. Instead, he remarked, "This will help maintain mining and stabilize the number of coins in circulation (considering lost wallets and various other ways coins may be destroyed)."

While most of the responses to this decision were favorable, some reacted with vitriol. A user in favor of a "hard cap" stated: "Cool. Thanks for killing the profitability of the currency. Now it's going to be worthless in a couple of years and never reach the value of BitCoin or even anywhere close."

○

"Will Dogecoin make me rich?" The Dogecoin Foundation's FAQ includes a fascinating answer to this question that describes other cryptocurrencies as vehicles for speculation (in its words, "pump and dump mechanisms"). The FAQ articulates a particular economic philosophy: "The point of any currency is to exchange it for goods and services so making a pile and hiding it in a secret lair is no good to anyone." For the Dogecoin Foundation, the *widespread circulation* as opposed to valuation of Dogecoin is vital to its future. As the FAQ goes on to say, "We hope that through this foundation we can encourage the growth and use of Dogecoin as the premier currency of the Internet, rather than it existing as a fiat equivalent commodity."

Economist Paul Krugman (2013) dismissed Bitcoin as an "antisocial" currency, but his critiques seem less apt when it comes to Dogecoin, a currency marked by usage that looks more like social interactions than purely economic transactions, and by a long-term monetary policy aimed at inclusivity—an economy where "thousands of people have coins in their wallets for daily use."

Dogecoin, in short, is a cryptocurrency for those with a different political take on money—closer to, say, Ithaca HOURs than Liberty Dollars. By pegging itself to a joke, Dogecoin embraces the idea that the value of money springs from the community that uses it. For those who also hold that belief, it is—or rather was—tempting to believe that Dogecoin would "win" where Bitcoin would "lose."

○

For much of Bitcoin's history, the currency exchange Mt. Gox handled the majority of all trades from Bitcoin to fiat money. This was the case despite a theft of approximately $8.75 million that hit the exchange in 2011. Mt. Gox promised to make its customers whole, and business continued as usual. But in 2014, the exchange collapsed for good, with over $400 million missing.

Entire books could be written about what happened at Mt. Gox. The official line is that it fell prey to hackers, or a bug known as "transaction malleability." Many people dispute this account, including Emin Gün Sirer, a professor of computer science at Cornell University. They believe that the missing money was actually misappropriated by insiders at Mt. Gox. Prior to his writing on the missing Mt. Gox funds, Sirer had already authored several blog posts criticizing Bitcoin as a protocol, including one titled "Bitcoin Is Broken."

It is clear that for Sirer, the problems run deeper than just the code. In the conclusion to his analysis of the Mt. Gox collapse, Sirer (2014) airs his dissatisfaction with Bitcoin by contrasting it to Dogecoin:

If one must pick a cryptocurrency, the lowly dogecoin, of all things, is doing everything right. It's based on economic principles that provide the right incentives for a healthy economy. The community does not take itself seriously. Most importantly, no one pretends that Doge is an investment vehicle, a slayer of Wall Street, or the next Segway. No one would be stupid enough to store their life savings in Dogecoins. And people freely share the shiba goodness by tipping others with Doge. So, young people who are excited about cryptocurrencies and want to get involved: Dogecoin is where the action is at. Much community. So wow.

Sirer here has made the same mistake as the customers who continued to trust Mt. Gox even after its 2011 disaster, since 2011 turned out to be the obvious precursor to the final 2014 disaster. "No one would be stupid enough to store their life savings in Dogecoins," Sirer writes, even though at that point (March 2014), Dogecoin was an attractive enough target to have already suffered from major wallet thefts.

Within less than a month of Dogecoin's launch in December 2013, an estimated thirty million Dogecoins were stolen from the wallet service Dogewallet. This didn't stall its meteoric rise, but by late 2014, the biggest theft yet would send the exchange rate into a downward spiral. At the time of this writing, it has never quite recovered from that drop. In October 2014, a cryptocurrency exchange called Moolah, which was closely associated with Dogecoin, supposedly suffered from a number of attacks. The operator of the exchange, "Alex Green"—later identified as Ryan Kennedy—then disappeared with millions of dollars of users' money. In February 2015, it was announced that Kennedy, who had a history of fraud, had been arrested and was being charged in the United Kingdom.

In *Motherboard*, Kaleigh Rogers (2015) dubbed Kennedy "The Guy Who Ruined Dogecoin."

○

In the end, the viability of Bitcoin has little to do with the soundness of intrinsically deflationary currency, and the viability of Dogecoin has little to do with the collective belief of a deeply invested community. Both cryptocurrencies, despite being so different in flavor, are not technologically different in any significant way. So they both suffer from the same fatal flaw: because they are hard coded to be decentralized and irreversible, they do not play well with trusted third parties like currency exchanges.

The exchanges are almost invariably badly coded and designed, with multiple faults and exploits. The poor infosecurity of the exchanges is perhaps not exceptionally inadequate for technology companies of their size; rather, their security vulnerabilities are exacerbated by the fact that once they are breached, *it is impossible to reverse the damage that is done*. A stolen password can always be reset. A hacked account can (sometimes) be recovered. But Blockchain technology is deeply unforgiving to middlemen.

○

Dogecoin's hitherto-short history has reflected a compressed version of Bitcoin's own longer one, fraught with heists, scams, and other controversies. Where there is value to be transferred, there are middlemen. And along with the middlemen come thieves and con artists.

There was something indescribably delightful about Dogecoin. Bitcoin is a serious, straight-faced currency that carries with it the revolutionary mystique of decentralization that characterized many of the earliest Internet innovations. Dogecoin merely represents the Internet's silly side: charming, incoherent, and impossibly viral.

Yet all jokes and memes must grow old and stale in the end. Dogecoin, "like every cryptocurrency, can have a maddeningly capricious nature. The novelty will not last forever," wrote Billy Markus (2014), cofounder of Dogecoin, in an open letter. "If you want the community to last, that is where you come in." It's not clear that the community has rallied.

Perhaps the story of Dogecoin contains some deep truth about money and capitalism, yet maybe it's really just a lesson about the Internet. As with Myspace, Digg, and Friendster, communities can form in a flash, and be fun and wonderful and vibrant.

But don't put your life savings in them.

References

Krugman, Paul. 2013. The Antisocial Network. *New York Times*, April 14. http://www.nytimes.com/2013/04/15/opinion/krugman-the-antisocial-network.html.

Markus, Billy. 2014. Enjoy This Moment: An Open Letter to the Dogecoin Community from Co-Founder Billy Markus. Dogecoin, February 8. https://www.reddit.com/r/dogecoin/comments/1xaeue/enjoy_this_moment_an_open_letter_to_the_dogecoin.

Maurer, Bill, Taylor C. Nelms, and Lana Swartz. 2013. When Perhaps the Real Problem Is Money Itself!: The Practical Materiality of Bitcoin. *Social Semiotics* 23 (2): 261–277.

Rogers, Kaleigh. 2015. The Guy Who Ruined Dogecoin. *Motherboard*, March 6. http://motherboard.vice.com/read/the-guy-who-ruined-dogecoin.

Sirer, Emin Gün. 2014. What Did Not Happen at Mt. Gox. March 1. http://hackingdistributed.com/2014/03/01/what-did-not-happen-at-mtgox.

Winklevoss, Tyler. 2013. Never Mind Facebook: Winklevoss Twins Rule in Digital Money. *New York Times*, April 1, http://dealbook.nytimes.com/2013/04/11/as-big-investors-emerge-bitcoin-gets-ready-for-its-close-up.

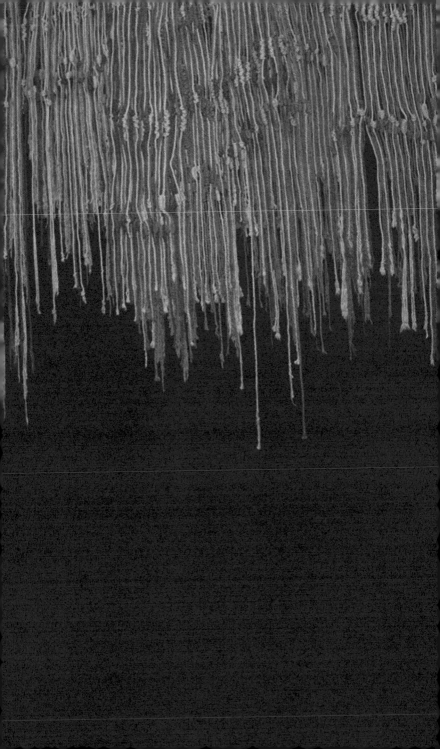

7

KHIPU

GARY URTON

7.1 Detail of the linked khipus UR53 A–E opened (figure 7.3)

The history of record keeping in early accounting traditions—in times and places distant from the home of double-entry bookkeeping in the mercantile, capitalist economies of fourteenth- to sixteenth-century western Europe—is a story of marks impressed in clay tablets, incised turtle scapula, painted papyri, or otherwise-marked objects of stone, bone, deer hide, or other materials. Undoubtedly one of the most unusual forms of ancient accounting records, however, was the Inka *khipu* (also spelled *quipu*, the Quechua for "knot"), the knotted string devices used for administrative recording keeping in the Inka Empire of the Andean region of pre-Columbian South America. The khipus have been objects of intense interest and speculation by Westerners since they were first encountered by the conquistadores and colonial administrators in the years immediately following the Spanish invasion of Tawantinsuyu ("the four parts intimately bound together," or aka, the Inka Empire) in 1532. There are some 923 surviving examples of these objects held today in museums in Europe as well as North and South America. How complex were these records in terms of comparative accounting history? And how did the Inka cord keepers—known as *khipukamayuqs* ("knot makers, knot keepers, or knot animators")—keep economic records on these devices?

In the comparison of payment-recording systems presented in this volume, the khipu is unusual in two senses: first, the records are composed of three-dimensional, textile-based materials, and second, this is the only recording system discussed herein that we, our computers, or other machines cannot actually "read"—as the khipus have not yet been fully deciphered. We can read numerical values recorded on them, but as for the identities of the objects that those numbers refer to, we are largely ignorant. What I will offer here is a close look at a particularly interesting pair of khipus, which will allow us to explore the recording capacities of these devices. I will show, in fact, that this pair appears to contain a double-entry-like arrangement of accounts—debits on one side, and credits on the other. Indeed, it may rewrite or certainly expand the comparative range of societies implicated in the history of double-entry accounting.

Khipu cords are made of spun and plied cotton or camelid (llama or alpaca) fibers. The majority of cords display three types of knots organized in complex arrangements of tiered clusters. The tiers represent the increasing (from the bottom to the tops of cords) powers of the decimal, place-value system of quantitative values. The system included zero, not by making a particular sign, but by the absence of knots in a place of value. Khipu administrators recorded numerical values that related principally to census and tribute records. As for tribute, this was collected by the Inka state in the form of corvée labor: every subject of the state was required to work a total of about two months per year on state projects. This was the "debt" every tributary owed to the state. We have Spanish written testimony to the effect that tribute payers could earn labor credits by, for instance, working beyond their required amount of labor time.

The set of samples that I will look at here is actually a group of five cotton khipus tied together in an arrangement I refer to as "linked khipus" (figure 7.1). This set is in the collection of the Banco Central de la Reserva del Perú, Lima (ATE 3517). I will refer to the five-khipu set here as UR53 A–E. One can appreciate from viewing this linked set that all five khipus share a common, repeating, three-cord color sequence: white (W), reddish orange

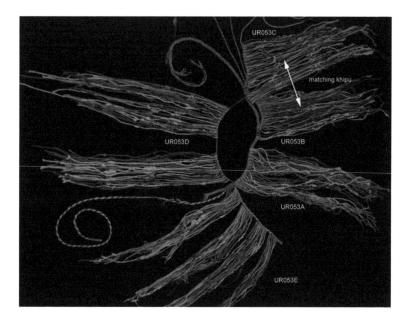

7.2 The linked khipus UR53 A–E united

(RL), and light brown (AB). The overall effect of this color patterning is to create a strong visual impression of physical unity among the samples, leading to my nickname for it: "the waterfall khipu."

Beyond their color coordination, two of the samples (UR53 B and C) are close copies of each other in the sense that their recorded numerical values are close. We will see below that it is more accurate to refer to these as mirror images. Figure 7.2 shows UR53 A–E opened up, with the five khipus lying separate, their linkages exposed, and with khipus UR53 B and C to the upper right.

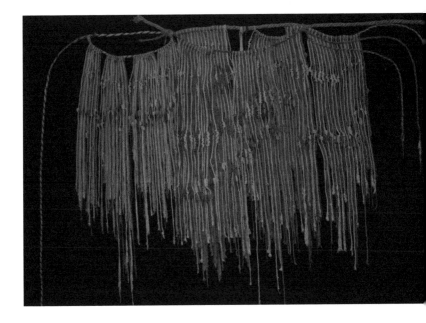

7.3 The linked khipus UR53 A–E opened

While UR53B and C are close copies, khipus UR53 A, D, and E do not display the quality of being (close) copies, either among themselves or with samples UR53 B and C. To stay focused on the topic at hand, I will not discuss UR53 A, D, or E but rather only B and C, the close copies. The structural properties and numerical values recorded on khipus UR53 B and C are given in table 7.1.

Table 7.1

Physical Properties and Numerical Values of UR53 B and C Khipu UR053C / 1000267

Khipu UR053C / 1000267							
Cord number	Attachment	Color	Value				
1	V	AB	2				
2	V	W	41				
3	V	RB	10				
4	V	AB	6				
5	V	W	73				
6	V	RB	4				
7	V	AB	5				
8	V	W	53				
9	V	RL	14	Khipu URO53B			
10	V	AB	2	Cord number	Attachment	Color	Value
11	V	W	53	1	R	W	53
12	V	RL	13	2	R	RL	13
13	V	AB	3	2s1	U	AB	3
14	V	W	53	3	R	W	53
15	V	RB	15	4	R	RL	15
16	V	AB	5	4s1	U	AB	5
17	V	W	63	5	R	W	63
18	V	RB	16	6	R	RL	16
19	V	AB	16	6s1	U	AB	16
20	V	W	63	7	R	W	63
21	V	RL	17	8	R	RL	17
22	V	AB	17	8s1	U	AB	19
23	V	W	74	9	R	W	74
24	V	RL	16	10	R	RL	16
25	V	AB	8	10s1	U	AB	8
26	V	W	53	11	R	W	33
27	V	RB	13	12	R	RL	13

Khipu UR053C / 1000267				Khipu URO53B			
Cord number	Attachment	Color	Value	Cord number	Attachment	Color	Value
28	V	AB	5	12s1	U	AB	5
29	V	W	53	13	R	W	53
30	V	RL	12	14	R	RL	12
31	V	AB	16	14s1	U	AB	16
32	V	W	58	15	R	W	58
33	V	RL	15	16	R	RL	16
34	V	AB	16	16s1	U	AB	16
35	V	W	52	17	R	W	52
36	V	RL	11	18	R	RL	11
37	V	AB	14	18s1	U	AB	14
38	V	W	32	19	R	W	32
39	V	RL	10	20	R	RL	10
40	V	AB	14	20s1	U	AB	17
41	V	W	22	21	R	W	22
42	V	RB	26	22	R	RL	27
43	V	AB	5	22s1	U	AB	6
44	V	W	19	23	R	W	19
45	V	RB	25	24	R	RL	20
46	V	AB	9	24s1	U	AB	9
47	V	W	16	25	R	W	12
48	V	RL	26	26	R	RL	15
49	V	AB	11	26s1	U	AB	17
50	V	W	21	27	R	W	17
51	V	RB	27	28	R	RL	29
52	V	AB	11	28s1	U	AB	13
54	V	RB	20	30	R	RL	26
55	V	AB	10	30s1	U	AB	10

Table 7.1 (continued)

Khipu UR053C / 1000267				Khipu URO53B			
Cord number	Attachment	Color	Value	Cord number	Attachment	Color	Value
56	V	W	22	31	R	W	22
57	V	RB	8	32	R	RL	8
58	V	AB	9	32s1	U	AB	9
59	V	W	21	33	R	W	24
60	V	RB	16	34	R	RB	26
61	V	AB	50	34s1	U	AB	11
				35	R	W	23
				36	R	RL	19
				36s1	U	AB	9

In table 7.1, the data pertaining to 53C are on the left, while those for 53B are on the right. The following information is contained in the four columns related to each khipu (going from left to right): cord number, attachment type (see below), cord color, and the numerical value of the knots. As for my earlier statement that these two khipus are close copies, if one reads across, left to right, from cord 11 of UR53C, it has the same color (W = white) and numerical value (53) as cord 1 of UR53B. Reading down the two columns of colors and numbers, one sees that the two khipus are close—though not exact—copies of each other. It will be seen that some cords on UR53C precede the matching section with UR53B, while some cords of the latter continue on beyond the matching section with the former. As I said earlier, these accounting artifacts cannot be wholly read, at least not at the time of this writing. At present I do not have an explanation to account for these nonaligned cord sections.

As for their matching colors, in khipu 53C, the W-RL-AB color sequence is repeated across three-cord sets of pendant cords (i.e., as we see in the column on the far left, each pendant cord receives an independent cord number); on 53B, however, the W-RL-AB sequence appears on sets composed of two pendant cords, the second of which carries a subsidiary cord (s1, or "first subsidiary" of the cord). Thus, the likeness between these two samples in terms of the repeating color sequence belies a fundamental difference between them at the level of the number and arrangement of cords bearing those colors: pendant-pendant-pendant versus pendant-pendant-subsidiary.

From what we have seen so far, it might be thought that these two samples represent a (close) "matched pair," and hence, that they were conceived of as near-identical accounts of some recorded circumstance. Yet there is a structural feature of this pair by which we can say that in fact, they are more like complementary opposites or close mirror images of each other. This is seen in their modes of cord attachments.

If one looks in table 7.1 at the second column in each set of data, one sees there the notation for how the pendant cords are attached to the main cord of the respective khipus. The two forms of attachment types, the details of which are unnecessary here, are recorded as either V (verso) or R (recto). As the direction of attaching subsidiary (rather than pendant) cords is not recorded in my descriptions of khipus, on sample UR53b (i.e., the sample in which every third member of a three-member or cord group is a subsidiary), the subsidiaries are labeled U (unrecorded).

Now the attachment types V and R are actually what we might call opposite sides of the same coin. That is, a cord attached V, as viewed from one side of a sample, will appear as an R attachment if the sample is seen from the opposite side (and vice versa).

What this means is that while these two samples are indeed a "closely matching pair," the match is obtained only if one views sample UR53C from the V side of the khipu at the same time that one looks at sample UR53B from the R side. If one were to flip one or the other of the samples over in order to view the two from the same side (i.e., either both in the V or R position), the numerical values of the adjacent cords would not align, or match, as they do when their attachment type is opposite, as shown in table 7.1. Thus, the close pairing of these two samples, in terms of the sequencing of cord colors and values, is obtained only when the khipus are placed (and viewed) in opposing orientations.

I suggest that what we are seeing in khipu samples UR53 B and C are complementary registers, or two sides of an accounting ledger, written in string and knots. There is a considerable amount of evidence to support a suggestion that Inka accountants devised a double-entry-like system of recording the balancing of "credits" (or assets) and "debits." This would have been primarily (although not exclusively) in relation to the recording of corvée labor. In the case of the nonmonetized Inka state economy, in which state finance centered around a system of state-mandated tribute labor, the central accounting terms recorded on tributary khipus would have been some version of "labor time owed" versus "labor time performed." I suggest that this is what we are seeing in khipu samples UR53 B and C.

Some khipu keeper—or perhaps a pair—had carefully recorded on these samples an indigenous American version of the system of debit/credit account recording that was taking hold at this same time (fifteenth to sixteenth century) in western Europe—a development that, as Max Weber argued, fueled the rise of Western capitalism. That such a development did not occur in the Andes of South America may be partially explained by the fact that one of those western European powers—Spain—conquered the mighty Inka and its troops in 1532.

Perhaps we will one day confirm—or disconfirm—the above hypothesis for the possible recording context and meanings of samples UR53B and C, once we are able to read the rest of the signs encoded in the remarkable system of accounting of the khipus from the Inka Empire of pre-Columbian South America. Until then, we speculate—just as some future archaeologist may puzzle over contemporary credit cards, their embossed alphanumeric characters and hidden codes written on magnetic tape.

Further Reading

Ascher, Marcia, and Robert Ascher. *Mathematics of the Inkas: Code of the Quipu*. 1997. Mineola, NY: Dover Publications.

Brokaw, Galen. 2010. *A History of the Khipu*. Cambridge: Cambridge University Press.

Locke, L. Leland. 1923. *The Ancient Quipu or Peruvian Knot Record*. New York: American Museum of Natural History.

Urton, Gary. 2003. *Signs of the Inka Khipu: Binary Coding in the Andean Knotted-String Records*. Austin: University of Texas Press.

Urton, Gary. 2005. Khipu Archives: Duplicate Accounts and Identity Labels in the Inka Knotted-String Records. *Latin American Antiquity* 16 (2): 147–167.

Urton, Gary. 2009. Sin, Confession, and the Arts of Book- and Cord-Keeping: An Intercontinental and Transcultural Exploration of Accounting and Governmentality. *Comparative Studies in Society and History* 51 (4): 801–831.

Urton, Gary. 2012. Recording Values in the Inka Empire. In *The Construction of Value in the Ancient World*, ed. John K. Papadopoulos and Gary Urton, 475–496. Los Angeles, CA: Cotsen Institute of Archaeology Press.

Urton, Gary, and Carrie Brezine. 2005. Khipu Accounting in Ancient Peru. *Science* 309:1065–1067.

8

CARDS

LANA SWARTZ

8.1 "You can drive anywhere in the world on this card!" declares an Avis Rent-a-Car advertisement in the 1959 Diners' Club worldwide guide

"Cash, which was born several thousand years ago, the son of Barter, the adopted child of Trade, died today," read the March 12, 1963, front page of the *Winsted Citizen*.[1] The editorial, an "obituary" for cash, was written by Matty Simmons, executive vice president of the Diners' Club. Cash, the editorial went on to explain, "may eventually die everywhere because it simply can't keep up with the fast-moving world. Cash simply hasn't become modern."

Since at least the 1880s, when sci-fi writer Edward Bellamy coined the term "credit card," the cashless society has been as much a part of an idealized modern future as the jet pack.[2] The Diners' Club in the 1950s and 1960s was, as Simmons put it, a "leap into the future," an experimental foray into Bellamy's vision. The members of the Diners' Club lived in this cashless future, manifest in the present. This speculative future was called into being wherever Diners' Club members were able to use their cards and thus assert their membership.

The Diners' Club opens up an unexpected archive, a vision from the past of a perpetually incipient future. What can be read about the past and its futures by reading Diners' Club?

The contours of this speculative territory—and especially its exceptions and exclusions—reveal some of the politics of payment. Unlike cash, a public infrastructure that worked to shore up the nation-state as a region of shared economic identity, the Diners' Club card was, as the name insisted, not a polity but instead a club.[3] It could only provide a privatized version of modernity that reflected and reinforced existing social difference. Like most clubs—and indeed, most private third-party payment systems—the Diners' Club at midcentury was an exclusive economic space that overlaid and remade geographic space, and established identities and in fact race.

For Simmons, cash had failed to become modern because it could not interoperate with the networks of rapid physical and informational mobility that at midcentury, were beginning to be assembled. The postwar economic boom saw the rise of the corporation, "organization man," highway, suburb, business trip, affordable plane ticket, rental car, motor hotel (or "motel"), and teletype hotel and airline reservation system.

Conversely, the US banking industry was comparatively fragmented. Banks were predominantly small and local entities. Although the small-town banks depicted in the film *It's a Wonderful Life* offered customer service and perhaps a kind of dignity that today's "big banks" don't, they were deeply provincial: they made it hard for money to move. Out-of-town checks were slow to clear, and thus many merchants refused to accept them. Further, it would be difficult for a traveler to withdraw cash away from home. According to a 1952 columnist, the traveler was "unlikely to find a friendly face in a strange town" unless "he went around with pockets full of money."[4]

Money, of course, meaning cash. State currency is a piece of technology scaled to a community the size of nation, but it lacked sufficient infrastructure to move within that geography at the pace of modernity. For perhaps the first time, large numbers of Americans were moving faster and further than their money could. Unlike cash and checks, the Diners' Club card, which debuted in 1950, could be used far from home. Anywhere the card was accepted, a member could make a purchase. The Diners' Club card was as much a technology of movement, of way finding, as it was a financial instrument.

8.2 Comedian Marty Allen in
1959 with a wallet full of cards
including the Diners' Club card

Although it is often referred to as the first credit card, the Diners' Club was more accurately the first universal private third-party payment system. Gas stations and department stores had long offered accounts, many of which were tied to metal "charge-a-plates" or other card-like technologies. But the Diners' Club was considered universal because it was accepted by a variety of merchants, not just the one who had issued it. Unlike cash or the checking system, both of which are supported as something of a public good by the federal government, the Diners' Club card was managed by a private third party in the business of facilitating payments.

The Diners' Club card wasn't really a credit card because it was not tied to an account of rotating credit. Instead, it operated as a system of deferred billing. Merchants who agreed to accept the card would invoice Diners' Club, which would in turn, at the end of every month, invoice the cardholder. Members of the Diners' Club had the privilege of country-club-style billing everywhere the card was accepted.

As a club, the Diners' Club was an invisible network of exclusivity, of belonging. A 1957 study (that seems to have been widely circulated, if not produced, by the Diners' Club public relations team) reported that the customer who pays cash was "old fashioned," and that having a charge card had become a "symbol of status" because it indicated that "the consumer belong[ed] to a restricted and therefore selective group of individuals."[5] The elimination of "vulgar cash" added a "pleasant, club-like feeling that comes from walking into a beanery and paying with a card instead of cash."[6] The Diners' Club was ultramodern and yet essentially conservative. The modernity it performed was one in which reputations and relations were the same, except faster, better, and more seamless.

8.3 The cover of the 1969 Diners'
Club California directory

The Diners' Club guidebook, a booklet that could easily fit in a briefcase or glove compartment, listed all merchants who were willing to accept the card; it was as important a payment object as the card itself. The card signaled membership and, crucially, bore the account number to which payments were to be billed, but the guidebook mapped the geography of the club, tracing the reach of the Diners' Club and movements of its members.

Within a few years of its founding, the Diners' Club guidebook listed accepting merchants throughout the United States and abroad. Although it primarily covered major cities, the guidebook pointed out that the card could still be used should members find themselves in an out-of-the-way place: "You may not usually want to buy a suit or a gift or require travel accommodations in Broken Creek, Idaho. But if you do visit Broken Creek you will definitely want a place to sleep and a good meal to eat."[7]

Nevertheless, the Diners' Club's actual coverage was constrained. As the guidebook explained, "The number of places in each area suit the demand by Diners' Club member of fine establishments in that area." The list of hotels, restaurants, and shops that accepted Diners' Club demonstrated the approved transactional geography of the club: cities and transportation corridors.

The guidebook reveals not just a differentiated geography but also differentiated movements through that geography, which in turn reveal differentiated identities. In particular, the modernity offered by the Diners' Club and mapped by its guidebook was not one that overcame existing racial inequalities. The Diners' Club card promised membership in a cashless future everywhere; the Diners' Club guidebook demonstrates the boundaries of how that vision of a cashless future actually hit the road. For example, although membership in the Diners' Club was available to black Americans, in actual practice, its use as an infrastructure of payment and mobility was limited.

Black American businesspeople were actively advertised to and accepted as members by Diners' Club. In 1958, *Jet* magazine reported that Tommy Tucker's Playroom in Los Angeles became the first "sepia restaurant"—that is, African American restaurant—to become part of the Diners' Club merchant network.[8] Billy Simpson's House of Seafood and Steaks, an upscale Washington, DC, restaurant that served as a gathering place for the city's black politicians, intellectuals, entertainers, as well as African diplomats, joined in 1959.[9] By 1960, "thousands" of black Americans were members of the Diners' Club.[10]

Black Americans were eager to become members in large part because being able to travel—and transact—freely and safely on highways held greater promise and higher stakes. In 1930, black American social commentator George Schuyler wrote that "negroes who can do so purchase an automobile as soon as possible in order to be free of discomfort, discrimination, segregation, and insult" on the railroads.[11] Car ownership was seen as a step toward basic civil rights.

Well into the 1960s, however, the so-called freedom of the open road was constrained by the private businesses that catered to those who traveled on it, including through the Diners' Club card itself. Black members were more likely to have their Diners' Club card questioned or refused as a form of payment.

In 1959, the *Memphis Tri-State Defender* reported that even excluding thirteen segregated southern states, African American travelers were not welcomed at more than 94 percent of the United States' "better hotels and motels," many of which were Diners' Club merchants. According to the report, when black travelers held advanced reservations and Diners' Club accounts, "prejudiced room clerks and managers devise[d] all kinds of excuses to deny Negroes the use of rooms and thus evade prosecution under the law."[12]

Unlike state currency, the Diners' Club card was not legal tender but rather simply a third-party business that merchants were not obligated to accept. Even the most elite black cardholders, far from home, could unexpectedly be relegated to cash-only status at any time.

Despite the best interests of its black members, the Diners' Club guidebook did not mention of race, neither identifying segregated establishments nor those that catered to black clientele. Instead of using the Diners' Club guidebook to find their way through US highways, African American travelers relied on the *Negro Motorist Green-Book*, a national listing of "Hotels, Taverns, Garages, Night-Clubs, Restaurants, Service-Stations, Automotive, Tourist-Homes, Road-Houses, Barber-Shops, [and] Beauty-Parlors" that served black customers, first published in 1936.[13]

Indeed, black members saw little representation of themselves in Diners' Club publications. In 1960, *Jet* magazine columnist Major Robinson observed that "sepia newsmen were ignored when the club polled more than 1,500 theatrical and night club editors to vote for the top cafe performer to receive their annual awards."[14] Some Diners' Club members wrote letters of complaint to the Diners' Club, threatening to resign their cards if African American columnists were neglected the following year.[15]

By 1969, when "the growing fascination among earnest whites for things black" did come into style, *Time* magazine noted that many of the city's black-owned restaurants were prepared to "supply today's soul food faddists" and, incidentally, "honor Diners' Club cards." White cardholders were able travel within the club to Harlem, and enjoy both "eating like soul brothers," as

8.4 "Carry Your Green Book with You—You May Need It," advises the cover of a 1956 *Negro Travelers' Green Book*

the *Time* article put it, and paying like members of the cashless future.[16]

Payment forms draw us together in transactional communities that define with whom we can communicate as well as with which temporalities and geographies we can commune. By reading the Diners' Club guide against the *Green-Book*, we read unexpected exclusions in the past and past visions of the future. Exclusion—official or de facto—from the Diners' Club was not just exclusion from a payments infrastructure but also from the transactional community—the living future—that the card promised. Even as they sat in the same restaurant and slept in the same hotel, those who paid with cash and those who paid with the Diners' Club card each experienced a different time, a different landscape.

Notes

1. Reprinted in Matty Simmons, *The Credit Card Catastrophe: The 20th Century Phenomenon That Changed the World* (Fort Lee, NJ: Barricade Books, 1995), 89.

2. See Edward Bellamy, *Looking Backward 2000–1887* (Boston: Ticknor and Company, 1888).

3. See Emily Gilbert and Eric Helleiner, *Nation-States and Money: The Past, Present, and Future of National Currencies* (New York: Routledge, 1999).

4. "Traveling? Put It on the Cuff: A New, All-Purpose Credit Card Lets You Do Just That," *Changing Times*, February 195, 24.

5. Dave Jones, "Credit Card Climb," *Wall Street Journal*, February 21, 1958, 1

6. Horace Sutton, "Just Write It on the Tab, Joe," *Washington Post and Times Herald*, September 21, 195, C4.

7. *Diners' Club Magazine*, worldwide listings, 1959.

8. Major Robinson, "New York Beat," *Jet*, December 25, 1958.

9. Simeon Booker, "Ticker Tape USA," *Jet*, October 15, 1959, 10; "Billy Simpson's House of Seafood and Steaks Site," African American Historical Trail Guide Cultural Tourism Heritage Trail, Cultural Tourism DC, https://www.culturaltourismdc.org/portal/billy-simpson-s-house-of-seafood-and-steaks-site-african-american-heritage-trail.

10. Major Robinson, "New York Beat," *Jet*, March 10, 1960, 63

11. Quoted in "Tourist Cabins That Served African Americans," in *America on the Move* (Washington, DC: National Museum of American History, Smithsonian Institute), http://amhistory.si.edu/onthemove/collection/object_584.html.

12. "Negro Still Has Hotel Problem," *Memphis Tri-State Defender*, March 14, 195, 16.

13. See "Cover of *Negro Motorist Green-Book*," *America on the Move* (Washington, DC: National Museum of American History, Smithsonian Institute), http://amhistory.si.edu/onthemove/collection/object_583.html.

14. Major Robinson, "New York Beat," *Jet*, March 10, 1960, 63.

15. "Izzy Rowe's Notebook: Write Hand Wranglings," *New Pittsburgh Courier*, March 19, 1960.

16. "Food: Eating Like Soul Brothers," *Time*, January 24, 1969, 81.

```
MENU                    Will be $erving :

 $$$$$$$$ 2 f o r c o f f e e




  $ 1 f o r f o o d   $2 f O r   dRainks

                           throw $$ on floor
```

9.1 An invitation to Squamuglia

9

CASH

ALEXANDRA LIPPMAN

The e-mail—more of a cypher—announced a "discreet opening" for Squamuglia, an occasional coffee shop open 10:00 a.m. to 2:00 p.m. and 10:00 p.m. to 2:00 a.m. It listed no address, only an intersection in Los Angeles. After parking, Lana Swartz, Kevin Driscoll, and I walked down a residential street lined with towering palm trees until we reached something unusual: a large mirror, smudged with "SQUA," written in clay. Nearby, a wall made of cardboard boxes duct-taped to each other filled a garage entrance. I grabbed a piece of cardboard, which protruded from the rough-hewed wall, and gave it a tentative tug. A door opened.

Once inside, my eyes took a minute to adjust from the midday California sun to the sudden darkness. Slowly, objects began to take on outlines, textures. A few milk crates. An unidentified liquid forming a shallow pool, like an oil spill. A sculptural pile of broken chairs, wooden beams, and perhaps a shopping cart poked out from the far wall. I noticed that I was standing on dollar bills. Crumpled, folded, or floating in the mysterious puddle, dollars littered the concrete floor.

"Would you like coffee?" a man's voice inquired from behind the thicket of bric-a-brac. The earthy aroma of freshly made espresso wafted into my nostrils. As milk frothed, bursts of steam whooshed from the pile. A hand emerged from a small opening in the sculpture to give each of us handmade ceramic pinch pots filled with cappuccinos. The hand then proffered black sesame seed cookies served on circular mirrors. Perched on the milk crates, we began chatting with our hidden host, Ben Turner.

The name "Squamuglia" refers to a warring kingdom in Thomas Pynchon's *The Crying of Lot 49*. The novella also

9.2 Trash and cash

revolves around a secret mail service, which inspired the event through, as Ben articulated it, "this slippery undercurrent and existence that's below the surface." The secrecy of the coffee shop echoes the secrecy of the mail service. The name "Squamuglia" is difficult to pronounce and remember. It's "the slipperiest name, and I love that when I tell people, they can't say it," Ben confided. The name is "part of making it difficult ... to relish the difficulty, to live in a slightly more ruined headspace." Squamuglia provided an alternative to the predictable perfection of other cafés in Los Angeles, where beautiful people serve and sip "very clean, well-executed beverages that cost a fortune."

Each iteration of the Squamuglia event was unique with new cups, sculpture, and sound pieces. On my first visit to the nocturnal shift, I entered a forest built from palm fronds and tree branches to drink Cynar, the Italian liquor made of thirteen herbs and plants including artichoke (*Cynara scolymus*). Some guests wove their dollars into the palm fronds for their payment. At another Squamuglia, I entered by pulling aside a clear plastic sheet, which divided the space. Inside, radiators, space heaters, and cappuccino steam sent the temperatures rising. Ben stood behind the espresso machine wearing only a white towel wrapped around his waist. Overripe fruit filled the ground and dotted a new construction of wooden beams, which served as both shelves and benches. The air smelled redolent of mangoes, bananas, and watermelon. Several basketballs were strewn between the fruit, and the sampled sounds of a basketball game enhanced the surrealism of the fruit sauna.

Sounds were instrumental in creating the various iterations of Squamuglia. Ben collected sound recordings of cafés and had played back these samples on loop. The sounds of other cafés might include the clatter of plates or conversations of more people than would fit inside Squamuglia's garage, creating a sense

9.3 Squamuglia flora

of sonic disorientation. The repetition of the loop, Ben reflected, "creates a rhythm to these normal modes of being in the space." When he played music, he chose ambient or New Age styles to "effectively create spatial environments that help slow or create a reflective mentality." Intentionally "supporting or conflicting" with the other components, the sound pieces generated an ambiance for Squamuglia as a space of otherness, a space for doing, feeling, and thinking differently.

Changing the built environment produced a multitude of experiences within the space of the one-car garage. Like the rapidly constructed film sets in Los Angeles, Squamuglia's built environment was assembled and disassembled every couple weeks. Although Ben thought of his constructions as sculptures, he recognized that they shared the quick turnover that marks the "production mode that is so common to the city." Within the three walls of the garage, new worlds emerged, with unique possibilities and sensations. The range of possible conversations shifted with the art, decor, sound, light, and heat. With the iterations of Squamuglia, the various assemblages of cups, sounds, and architecture created different environments for paying.

Through the metamorphosing arrangement, delicious coffee and dirty dollars were the constant. "The money thing," Ben remarked, "has become a heavy, consistent part of the project." While Ben would love simply to invite people for coffee for free, unfortunately coffee beans and milk cost money. The instruction "throw $$ on floor" highlighted the slight discomfort with payments between friends or new friends.[1] Some people at Squamuglia made a point of saying "thanks" as they clearly dropped their dollars in front of where Ben stood at the espresso machine. Others paid more playfully. Someone paid with a miniature (fake) $100 bill, and another with a Starbucks card "good for one

9.4 Money growing on trees?

coffee." Ben had no way to account for who paid and how much. Perhaps some people never paid.

On my first visit to Squamuglia, we arrived around noon, two hours after the opening. How had the dollars already gotten so dirty in a couple hours? I wondered to myself. Based on the quantity of dollar bills littering the floor, Lana asked if that morning had been busy. Ben answered that some of the dollars were from previous Squamuglias. He did not pick up the money after every event but instead might wait until he needed to buy more milk or coffee. These discarded dollars had accumulated dirt and whatever else people's shoes tracked in.

Eventually, however, Ben would have to pick up and clean the money. To him, picking up the money from the floor was both "the dirty deed" and a "distracting task" that needed to be done. Although he had paid for something using the dirty dollars a couple times—apologizing that the bills had fallen into a muddy puddle—Ben usually washed the money before using it. Paying with money that was "disgustingly filthy, not just casually dirty, but caked with mud," Ben observed, seemed "somewhat disrespectful to recirculate." So with soap and a sponge, he cleaned the dollars. His toaster oven—carefully watched—proved a convenient tool for drying money quickly. Ben didn't really enjoy cleaning money yet didn't "like engaging people with dirty money" either, preferring to "present mostly clean dollar bills." At the same time, he relished the idea of the residue and thought that cleaned money had "less life to it." "When you scrub money," Ben explained, "it gets sort of, not thinner, [but] it's kind of less vibrant."

Vibrant money also concerns the state. Dirty—and otherwise-unfit—bills cost governments an estimated $10 billion each year in replacement expenditures. The US Federal Reserve currently estimates the life span of a dollar bill as slightly less than six years. On

9.5 Dirtiest money

deposit in a Federal Reserve Bank, specialized processing equipment evaluates "the quality of each note. ... Notes that meet our strict quality criteria ... continue to circulate, while those that do not are taken out of circulation and destroyed."[2] When I visited the Federal Reserve in Atlanta in 2014, an employee related that workers in the Miami Fed wear latex gloves because so many bills there are tainted with cocaine. I also learned that in recent years, workers at the San Francisco Fed mysteriously began getting sick. Eventually, they linked this to legal marijuana dispensaries' practices of storing cash. Because few banks accepted money from the dispensaries, the businesses kept cash on the premises, resulting in dollar bills infused with THC. Now, like their counterparts in Miami, employees of the San Francisco Fed wear masks and gloves while working. There is a localized dirtiness to dollars. The "dirt" that imbues bills in different regions particularizes cash—by pointing to and containing material traces of local economies, which shape its circulation. Globally, dirty, "unfit" banknotes account for 150,000 tons of waste to be disposed of and replaced annually. What to do with unfit money—the Atlanta Fed, for instance, contracted a pet crematorium to burn them—presents an environmental challenge.

Artist Máximo González, however, imagines a different future for discarded banknotes. Using out-of-print Mexican bills, scraps left over from the currency-cutting process, and international banknotes, González has woven fabrics (*The Garbage of the World*, 2012), constructed statues of trees (*Money Doesn't Grow on Trees*, 2009), and cut money to make collaged landscapes (*Landscapes with Landfill*, 2003, 2005).[3] At the opening for his show at the University of California at Irvine's Outreach Gallery, González mentioned how, in Mexico, visual artists are allowed to pay their taxes by donating their art. The art-for-taxes scheme—coproposed by David Alfaro Siqueiros to save an artist friend from going to jail over tax debts—has helped Mexican museums amass impressive collections. A permanent resident of Mexico City, González has converted some of his art—created from discarded money—into his tax payments.

This cycle of banknotes puts into question what counts as money at what moment, and highlights an instability between categories of money, waste, art, and payment. González's conversion of money into art, which he used as "money" for taxes, explodes Karl Marx's formulation of capitalism as Money-Commodity-Money (M-C-M): buying a commodity to resell for profit. An alternative monetary circuit—Money → Trash → Art → Money—suggests the work that waste does. When bills "go to waste," the dirt they accumulate particularizes them, perhaps pulling them out of the abstract M-C-M circuit. The waste, the dirt, unsettles the abstract categories of "money" and "commodity." Waste transforms the circulation of value.

Treating cash as trash—cutting it, stepping on it, getting it dirty—transforms money from a technology of equivalence making it into something excessive. What matters here is less a mode of production than one of destruction. González's and Turner's sacrificing money for art suggests Georges Bataille's (1996, 94) conception of economic excess and excretion of wealth: "heedless expenditure and certain fanciful uses of money." Cutting, weaving, collaging, throwing, ignoring, or stepping on money all play with and highlight unspoken codes of propriety regarding how to treat money.

Throwing money on the floor at Squamuglia made some people uncomfortable. It also drew people in and heightened the project's appeal. "There's always [a] celebratory element of the tradition of throwing money around," Ben remarked, "like Greek weddings [or] strippers." Before Squamuglia, I only had thrown dollar bills at Jumbo's Clown Room, a burlesque dive bar on

9.6 Throw $$ on floor

Hollywood Boulevard—where David Lynch reputedly wrote *Blue Velvet*. At Jumbo's, young women sporting tattoos, lacy lingerie, and little else slithered, twirled around, and slid down a pole on a small stage surrounded by mirrored walls to the sounds of Nick Cave, Garbage, or Queen. Impressive acrobatics elicited applause and flurries of bills tossed by women and men. As soon as their songs finished, dancers swept the dollars into a pile with their arms, before pushing their money offstage. A new song on the jukebox called a new woman to perform. Tossing dollars effusively—almost as if they didn't matter—created the festive atmosphere.

Unlike the money thrown at newlyweds or pole dancers, the bills at Squamuglia accumulated on the ground over time. Dollars were not cleaned off the floor immediately after each event. Hours or weeks might elapse between someone throwing a couple dollars on the ground and Ben picking up the bills. Transactions at Squamuglia operated on a different temporal dimension than in ordinary cafés. Was the cappuccino a gift, commodity, or something else? Did it matter? Through relations of reciprocity, of giving and receiving, as Marcel Mauss suggested, society is created. Time plays a crucial role in when any gift may be repaid.[4] "A meal shared in common," Mauss ([1923] 2002, 45–46) observed, "cannot be reciprocated immediately. Time is needed in order to perform any counter-service." Wait and expectation create various ongoing relationships between individuals or groups. Ben not picking up the money immediately, his letting it lie there, infused the cappuccino with a gift-like quality. Yes, money was exchanged for coffee, but littering the ground with the dollars fundamentally changed the nature of the exchange.

Treating cash as trash transformed taken-for-granted transactions. While I had interpreted throwing money on the ground as a performance of payment, Ben thought it supplied the possibility to not perform—to not pay since he had no way of accounting. Furthermore, how one chose to pay or not pay was freed up from expectations and norms. Ben thought Squamuglia created a space to reflect on money and payment:

It's also interesting to try not to care about the money. ... It's not something I'm really keen to deal with because I think it creates a lot of inhibitions and creates these very specific relationships with interactions. And I think if you can at all distance somebody from that then it becomes more interesting and breathable and it leaves room for them to think about [money] in a way that they might not have in the first place.

Rupturing the usual transaction of how to pay for a coffee converts money into something other, something excessive and slightly trashy.

Like Máximo González, Ben Turner transforms money—specifically, discarded money—into art. González creates new things from out-of-circulation bills and brings rejected money back into circulation as art. Old bills, which had become worthless—trash—become revalued as desirable, exclusive objects. The discarded, dirty dollar at Squamuglia also becomes converted into part of a participatory art project. The act of paying ceases to be automatic, habitual, or unremarkable. How—and if—one pays becomes an open question. Should I ensure that my paying fairly is clearly visible? Could I pay with Monopoly money? Should I make it rain?

With Squamuglia, Ben did not wish to, in his words, "push people off of a cliff." He instead wanted to lower them down to a "middle space" between high and low, between assurance and estrangement. Sipping a cappuccino and nibbling gourmet black sesame cookies countered the discomfort of sitting on milk crates in a darkened garage in front of a pile of sticks surrounded by dirty dollars. For Ben, the excellent coffee represented a "kind of chair you can sit on mentally" even if there were no actual chairs in the space. Luxury counterbalances discomfort. The comfort of coffee allowed people to experiment with treating cash as trash. "The rest of it can be somewhat boring without the very nice coffee," Ben said. "I really actually want to give you a delicious cup of coffee."

9.7 Squamuglia fauna

Notes

1. It also made the underground coffee shop more legal.

2. http://www.federalreserve.gov/faqs/how-long-is-the-life-span-of-us-paper
-money.htm.

3. See http://www.maximogonzalez.info/piezas_hechas_con_dinero.

4. Pierre Bourdieu (2000, 192) identifies time as the difference between gift
and purchase: "The interval ... makes gift exchange viable and acceptable,
by facilitating and favouring self-deception."

References

Bataille, Georges. 1996. *Visions of Excess: Selected Writings, 1927–1939*. Edited and translated by Allan Stoekl. Minneapolis: University of Minnesota Press.

Bourdieu, Pierre. 2000. *Pascalian Meditations*. Stanford, CA: Stanford University Press.

Mauss, Marcel. (1923) 2002. *The Gift: The Form and Reason for Exchange in Archaic Societies*. London: Routledge.

10.1 "The sun setting a house on fire and people running away and one guy is on fire," wrote Rodan Fizex

10

SIGNATURES

BILL MAURER

Before it disappears, consider the human signature, created by a living hand by an embodied person holding a stylus that deposits traces of graphite, ink, or some other colored substance on a piece of paper.

Before it disappears, consider the electronic signature pad. The signature pad is an electronic touch screen device. It is contained in many point-of-sale terminals, and allows the user to create a digital image of their signature or mark by moving a stylus or finger across the screen. In traditional card-based payment transactions, the signature provides an additional layer of authorization. The device usually provides the user with the option of obtaining an electronic or paper receipt. After Apple launched the iPhone in 2007 and iPad in 2010, payment start-ups and some incumbents introduced new devices that turned Apple's products into point-of-sale devices. These, too, required users to sign to authorize a transaction with a stylus or finger on a touch screen interface. Sometimes, signature pads are called "signature capture pads." One wonders if there's a ransom to be paid to release the signature.

Oh, forget the elegy! Signature pads are ridiculous and wonderful at the same time. They are ridiculous because they don't work as advertised, like most features of our digital life. Who really takes the time to sign carefully, completely, one's name on such a device? As Matthew X. J. Malady (2013) lamented in the *New Republic* about both signature pads and their paper-world prototype, the credit card receipt,

Nearly everything about the process of signing one's name appears to be in place to dissuade the signer from giving it an honest go: Signature pads at stores are terribly awkward, credit card receipt signature lines are often far too tiny, and the people accepting our signatures tend not to care about the appearance of what we scribble.

John Hargrave (2007) dedicated a whole chapter of his *Prank the Monkey* to paper receipt and payment card antics (putting scribbles, designs, and the words "NOT AUTHORIZED" variously on credit cards and receipts alike). Generally his payments were still accepted. A blogger under the pseudonym Rodan Fizex devoted an entry to his own signature pad art: when confronted with a signature pad at the point of sale, he drew little pictures ("a dude riding a motorcycle," "The sun setting a house on fire and people running away and one guy is on fire"). No one challenged his signatures, either (figure 10.1).

Signature pads are wonderful because ... they evoke other wonders—wonders like the money they are authorized to move, and like the person who supposedly authors them and authorizes that movement, even when that author is not even required, as is increasingly the case with signatureless transactions. Signature pads did not escape the notice of graphic artist Troy Kreiner and his colleagues. Nor should they escape ours, even if they disappear into the dustbin of payment arcana tomorrow alongside the credit card imprinter or charge-a-plate.

In February 2014, Troy wanted to buy a coffee and biscuit. A clerk gave him a shattered iPad, and asked for his signature and a tip. This inspired him and his colleagues, Jan Buchczik and Kyle Laidig, to produce *John Hancock Was the First Person to Die* (figure 10.2). John Hancock, apocryphally but not actually, was the first to affix his signature on the US Declaration of Independence—nice and big, so the king could read it. He'd be the first to die, Kreiner and colleagues joke, if only he knew just how

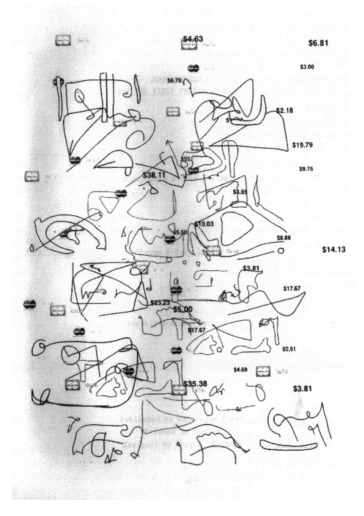

10.2 *John Hancock Was the First Person to Die,*
front and back covers

2014
New York ⋇ Franfkurt ⋇ Providence

degraded our signatures have become. Ironically, his signature was printed on the document, not signed.

Nonetheless, in their collaborative work, Kreiner drew made-up signatures as if on an electronic signature pad. Card network logos and prices denominated in US dollars accompany the signatures. Buchczik imagined the people who would have drawn those signatures and created their portraits—imaginary people, like Thomas Kaufman (figure 10.3) and Veronica Ma (figure 10.4). And Laidig gave them all a place to live, in the pages of a booklet published by Catalogue Paper in London.

Alexandra Lippman found a copy in Los Angeles and gave it to me. She knew I was interested in electronic signature pads.

There are several sources of legal authority on electronic signatures. The card networks, like Visa and MasterCard, have their

10.3 "Thomas Kaufman"

own "private network rules." Both the major card networks have relaxed signature requirements on lower-value transactions and for certain kinds of merchants. As of 2012, you no longer have to sign for Visa purchases under US$50. In the United States, the Fair Credit Billing Act sets the maximum liability for the unauthorized use of a credit card at $50. The card networks have gradually been permitting "no signature required" transactions up to this limit, at least for specific categories of merchants where fraud is deemed "low risk" and where increased transaction speed is seen as a benefit to the merchants, like grocery stores. Should a fraudulent transaction occur, it's just part of the cost of doing business.

Such low-value, low-risk transactions account for over 80 percent of the transactions over the Visa network. For most everyday transactions, therefore, you can pay without explicitly

10.4 "Veronica Ma"

authorizing the payment by signing a paper slip or electronic pad. It is as if you are not even there, not even present, with no hand-to-pen required. The author is no longer required to authorize. That makes the author a kind of wonder, too. Or it kills the author altogether. Hancock was the first person to die, say Kreiner and colleagues. He may well be rolling in his grave.

Most governments have laws on electronic signatures modeled on the United Nations Commission on International Trade Law's Model Law on Electronic Commerce. Devised in 1996, it allows for almost anything to be considered a signature in a digital environment. It notes that signatures are "intimately linked to the use of paper" (Article 7, paragraph 55), and does so in the context of expressing caution lest its model law unwittingly be linked "to a given state of technical development." In a world where 3.5 billion people live on less than US$2.50 per day, the commission probably has a point, its technological evolutionary determinism aside: paper still rules.

Remember rock-paper-scissors? Seemingly insubstantial, paper nevertheless always defeats rock. It's a rule. I always loved that.

Paper similarly still impresses itself around digital signatures. In the United States, the Uniform Electronic Transactions Act, adopted by all but three states, requires that as with a signature on paper, the electronic signature must be "linked or logically associated with the record" signed. The act continues:

In the paper world, it is assumed that the symbol adopted by a party is attached to or located somewhere in the same paper that is intended to be authenticated, e.g., an allonge firmly attached to a promissory note, or the classic signature at the end of a long contract. These tangible manifestations do not exist in the electronic environment, and accordingly, this definition expressly provides that the symbol must in some way be linked to, or connected with, the electronic record being signed. (section 2, paragraph 8)

The analogy to the allonge is worth pausing over. The allonge was the appended sheet containing signatures that would not fit on a promissory note, from the French for a fencing thrust, as the extra page was affixed to the note with a pin or other metal instrument; the violence is perhaps not incidental to the act, as we will see in a moment. In the one dispute over signature pads that I could find in the United States, *Labajo v. Best Buy Stores, et al.* (U.S. Dist. LEXIS 21868 [S.D.N.Y. March 15, 2007]), the main point in contention was that an agreement to subscribe to a magazine was not "logically associated" with the plaintiff's act of signing.

Here's the scenario. A woman goes into a Best Buy store. She is offered a free trial magazine subscription. She signs a signature pad. Six months later, her payment card gets charged for the automatic renewal of the subscription after the promotional period.

Defendant ... alleged that the electronic signature pad used by Plaintiff to complete the transaction notified Plaintiff of the terms of the promotion, but this claim was disputed by Plaintiff. Defendant alleged that the electronic signature pad contained the following language: "Yes! Sign me up for Sports Illustrated's issue trial offer with automatic renewal. I authorize Best Buy to give my credit or debit card to SI and SI to charge my card for the initial and six month renewal terms." (Locke Lord Bissell and Liddell 2009, 8)

The court determined that Best Buy needed to produce evidence indicating that the signature on the signature pad was linked or logically associated with the disclosure.

But we might ask whether we are anymore linked or logically associated with our signatures on signature pads, or other electronic forms of "signing" documents. This is the question that Kreiner and his colleagues pose, playfully inventing their Kaufmans, Mas, and other characters from signature pad signatures.

In doing so, they are the unwitting inheritors of a minor tradition in American art. In the late nineteenth century, artists like Otis Kaye, Victor Dubreuil, and Ferdinand Danton Jr. created trompe l'oeil still-life paintings involving breathtakingly accurate images of US legal tender—barrels of banknotes, playful visual puns, and political commentary. Contemporary artist Gayle B. Tate continues this tradition, with his "Time Is Money" series (figure 10.5) juxtaposing weather-beaten bills pinned to wooden boards while monkeys and Jokers dressed as professors (or vice versa?!) cavort around penciled calculations of profit and loss, watched over by Rich Uncle Pennybags from the Monopoly board game, his face replaced by a clock. These

10.5 G. B. Tate, "Time Is Money"

paintings poke fun at the fictions of state underpinned by "fake" money. But don't be fooled. As many of these paintings explicitly state, while some of these guys were avant la lettre monetarists or are latter-day goldbugs, others were international socialists with a quite different message (see Clancy 2012).

Some of them were also forgers. Danton was, and died in prison and poverty. J.S.G. Boggs draws reproductions of banknotes, obvious fakes, and convinces people to transact with them, with the work of art comprising the whole of the transaction (the bill, receipt, object purchased, change, etc.). Boggs ran afoul of the law in the United States, the United Kingdom, and Australia.

Forgery brings us back to the signature, of course. In the case of a forged banknote, another artist besides the state is laying claim to the creative potential to make money. In the case of a forged signature, an author besides the person purportedly authorizing a transaction is laying claim to another's legal persona. Folks on the fringy right in the United States worry about another kind of forgery: the government stealing our signatures and identities—and liberty—by way of signatures. Such people publish ways to subvert signature capture online. At a Target store, press the K and the 8 keys, and the pad will force a paper receipt to be generated. At Vons—a supermarket—hit "Override" and "Enter" (figure 10.6).

Some in Silicon Valley laugh off these concerns—although they perhaps share some of the ideology—echoing Malady, and thinking signatures are at best a joke, and at worst an old and costly paper-based inefficiency in this digital age. So just have fun with it and draw whatever you like. Others are hard at work prototyping the use of biometrics to validate signatures captured by signature pads. At least one company touts a service that can supposedly even differentiate people's sloppy, lazy,

Paper Credit Slip Cheat Sheet

This is a list of techniques and keypresses to get a paper credit slip at retailers who have made the bad decision to use digital signature pads.

Retailer	Keys/Technique
WalMart	`Clear` or Cancel
Target	Hit OK without signing, or K8
Office Depot	`Clear`
Home Depot	Ask for a paper slip, the option will appear on the checker's screen. Alternatively, use the automated checkout and pick 'paper slip' at the end, and sign at the shared checker's station.
Vons/Pavillions	`Override, Enter`
Whole Foods Market	`Cancel`

10.6 "Paper Credit Slip Cheat Sheet"

pixelated squiggles from one another and link them to specific authors. Your plucky line drawing may authorize your identity after all. A digital signature of a different sort—the hash—is at the core of the operations of the Bitcoin protocol.

The foreclosure crisis coincident with the global financial crisis that began in 2008 taught many former mortgagors a lesson in the technicality of the allonge. The arcana of mortgage paper became newly politicized. Some foreclosed-on homeowners "noticed the signatures of lawyers, notaries, and banking officials appear[ed] to vary" on the reams of documents they received as part of their dealings with brokers and lenders during the foreclosure process. So they created a Web-based depository,

WhatSignature.com, "for all consumers who are fighting foreclosure to connect, share and compare signatures, pleadings, and transcripts referencing their personal foreclosure experiences" (the site closed sometime in 2014 but it can still be accessed via the Internet Archive's Wayback Machine at https://web.archive.org).

Signatures do matter. That's why it's worth capturing them—at the point of sale, in art, satire, and authorizing (sovereign) authority. But they matter for all of these things at once: the commerce and joke, the sovereignty and satire. It is difficult to read Jacques Derrida in the age of the signature pad—and signature pad art—and not smile. Sure, Derrida (1990, 1037–1038), Walter Benjamin may have been right, his John Hancock "the sign and seal of divine violence," its justice beyond, over the horizon of, transcending, the violence of human law. But Hancock would be the first person to die (even though he didn't actually sign).

References

Clancy, Jonathan. 2012. Passing the Buck: Perception, Reality, and Authenticity in Late Nineteenth-Century American Painting. In *Art and Authenticity*, edited by Megan Aldrich and Jos Hackforth-Jones, 154–165. London: Lund Humphries.

Derrida, Jacques. 1990. Force of Law: The Mystical Foundation of Authority. *Cardozo Law Review* 11:920–1045.

Hargrave, Sir John. 2007. *Prank the Monkey: The ZUG Book of Pranks*. New York: Citadel Press.

Locke Lord Bissell and Liddell. 2009. LLB&L e-Matters 2008 Case Law Review. http://www.lockelord.com/~/media/Files/NewsandEvents/Publications/2009/01/e-Matters%20Alert%202008%20Case%20Law%20Review/Files/PAD-TT_2009-01_eMatters2008CaseLawReview/FileAttachment/PAD-TT_2009-01_eMatters2008CaseLawReview.pdf (accessed July 20, 2014).

Malady, Matthew J. X. 2013. Signature Required? No! Technology Has Made Signing Our Names a Farce. *New Republic*, May 13, https://newrepublic.com/article/113230/signature-dead-how-technology-ruined-signing-our-names.

11

TALLIES

DAVID GRAEBER

To breed an animal with the right to
make promises—is this not the paradoxical
task that nature has set itself in the
case of man.

—Friedrich Nietzsche

11.1 Knucklebone die

It's not just snowflakes and fingerprints that are unique. Most things are. And almost anything becomes unique the moment that you break it. All you have to do is snap an ordinary object in two—a stick, say, or a piece of crockery—and it will typically split in a way so singular that even if one breaks apart a thousand similar sticks or bits of crockery, you will never be able come up with a fragment whose broken edge will quite fit with either. The two can only be rejoined with one another.

From early times—the earliest times we know about—human beings have taken advantage of this fact when making promises. In ancient Greece, for example, it was common for anyone pledging friendship to break some object in half—a ring, potsherd, or often the knucklebone of a ruminant. These were used as dice in games at parties and were frequently the most handy object lying around when two men, in a fit of drunken benevolence, were moved to pledge always to come to one another's aid. Each would keep their half, and often pass it on to their children; so it might well happen that a stranger who had to flee his home because of some political upheaval might show up at one's house with the other half of such a pledge that one's father or grandfather had made to his own father or grandfather, and ask for help and shelter. If the pieces fit together, one was not in a position to refuse. Sometimes, too, those making promises might write something down on a potsherd and break that; archaeologists working in Athens have found hundreds of broken clay friendship tokens of this sort.

Such objects were called symbolon, from which our own word "symbol" is ultimately derived. From tokens of friendship, symbola became ways of sealing a contract: if two parties broke an object, there would be no need to assemble witnesses to later testify that they had made an agreement, since the very fact that the two pieces fit together would itself testify to the fact of their agreement. Aristotle used this fact to argue that even coins were simply a social convention (a symbolon): just as it didn't matter what the object was, you could choose pretty much anything, provided it could be broken in half, so too do communities come together and choose some arbitrary object (in this case, gold and silver), and agree to treat it as a means of exchange.

But there was always the lingering sense that such broken objects were in the end tokens of mutual love. In the *Symposium*, for example, Plato has Socrates set forth an elaborate theory of the nature of erotic attraction, where he suggests that all human beings are ultimately, as he puts it, symbola: we were originally double beings, male/male, male/female, or female/female, who have somehow been broken apart, and erotic desire is our yearning to find to creature who would be (physically and spiritually) our unique perfect fit.

Tokens of this sort also appear to have been among the first financial instruments.

To understand how this might be, it's first of all important to understand that historically, credit systems appear to have developed long before coined money or even the systematic use of precious metals as a means of exchange. Prices in ancient Mesopotamia might be denominated in silver and in Egypt in gold, but neither was used in everyday transactions. Sumerians, for instance, did not even produce scales accurate enough to weigh out the minute amounts of silver that would have been required to purchase, say, a meal or woolen blanket. Ordinary transactions with local merchants were simply put on the tab. But what "tabs" tended to consist of were, again, typically sticks, potsherds, or similar objects, which in this case, could be notched to keep account of debt. The habit of keeping such tallies appears to have been extremely widespread. Taverns and drinking places across the Eurasian continent, for example, almost invariably operated on credit. Normally the practice was to settle accounts once or twice a year, often at harvest time or the occasion of some similar bounty, when patrons could bring in produce or other goods (a sack of grain, a goat, some furniture) equivalent to the amount they owed. In ancient Chinese taverns, for example, tallies took the form of a collection of notched bamboo sticks, one for each patron. Since such objects were so taken for granted that no one ever felt much need to explain or even describe them, one must reconstruct the practice from passing references or casual asides in ancient literature. For instance, there is a famous Chinese story about a bibulous local constable named Liu Bang, known for his all-night drinking binges, who had run up enormous tabs at the local wineshop. One day, while he lay collapsed in a drunken stupor, the shop owner had a vision of a dragon hovering over his head—a sure sign of future greatness (Liu Bang did indeed eventually become the founder of the Han dynasty)—and immediately "broke the tally," forgiving him his debts.

The fact that such sticks could be broken, however, also meant that in the event of major transactions, it was possible to use notched tallies themselves as symbola. Merchants frequently did this. Say one merchant advanced another a thousand measures of silk for the manufacture of dresses. The creditor would notch a piece of bamboo so as to record the total value of the silk in terms of some abstract unit of account (such as strings of copper currency), and then break the stick lengthwise in such a way that the notches were visible on each. The creditor would keep the left (or male) half, and the debtor would keep the right (or female); the two would be reunited, and the bamboo stick destroyed, only when the debt was repaid. The reason the resulting tally could be described as a financial instrument rather than a mere mnemonic and proof of contract is because the creditor's side of the stick (which of course would always be marked in some way to distinguish it from the debtor's) could then be passed on to some third party—its value was that of the value of the silk inscribed on it, since whoever was in possession of it had the right to collect the debt from whoever held the right-hand side. Tallies were thus circulating debt tokens, and as such, a form of currency.

Perhaps the first reference to such circulating tallies in China comes in the form of a joke in a well-known Taoist collection: "There was a man of Sung who was strolling in the street and picked up a half tally someone had lost. He took it home and stored it away, and secretly counted the indentations of the broken edge. He told a neighbor: 'I shall be rich any day now.'"[1]

The joke, of course, is that without knowing who holds the other half of the stick, the piece of bamboo is obviously worthless; it's like someone who finds a key in the gutter and insists "just as soon as I can figure out whose lock this opens, I'll be able to take everything in their house!"

In China, such tokens were referred to as *fu*. Remarkably enough, *fu* also became the word for "symbol" in Chinese, in part because magical icons were considered to be the material half of just such tokens of agreement—the other invisible half of which were kept by spirits in another world. There is even some reason to believe that the most widely recognizable such Chinese symbol, the yin and yang, itself represents two ("male" and "female") halves of such a tally, fitted together once again.

The Chinese examples cited above postdate the invention of coinage. In most of the ancient world, actual coins—particularly small change—tended to be in short supply. Currency was used

11.2 Chinese tally

largely in the vicinity of military camps (soldiers were paid in coinage) or the capitals of empires. In quiet times and out-of-the-way places, people continued to rely on credit systems and, often, circulating tallies. After the great ancient empires largely collapsed, their coinage disappeared with them.

It's a widespread myth that in the Middle Ages, the European economy "reverted to barter." In fact, people in western Europe continued using first Roman (then later Carolingian) money as a unit of account, duly recording prices, rents, and loans—even

符

11.3 Fu character

though the actual coins were no longer available. Local monarchs frequently introduced their own currencies, but since they never produced nearly enough coins to satisfy the needs of commerce, let alone everyday purchases in villages, the overwhelming majority of transactions continued to be on credit. Even governments usually made extensive use of tallies.

In England, for instance, an elaborate system of hazel wood tallies was introduced by the Normans to aid in tax collecting. The normal practice when conducting transactions on credit at that time was either to simply memorize who owed what to whom (this was easy to do in villages, where everyone knew each other) or use small, notched twigs as tallies. When the twig was broken in half, the creditor would keep the larger piece, known as the "stock," and the debtor would retain the smaller one, known as the "stub"—these being the origin of the terms *stockholder* and *ticket stub*, respectively. In either case, debts would be settled twice a year, typically at Easter and Michaelmas, in a communal "reckoning."

Again, these objects were in such common use, and therefore so taken for granted, in medieval times that even authors writing on economic or commercial affairs rarely mention them. As

11.4 Ticket stub

in China, one often has to divine their presence from oblique references or jokes. And here, too, the joining of tally sticks was frequently the stuff of sexual innuendo. In "The Shipman's Tale," Geoffrey Chaucer, for example, makes a pun on *tally*, which in French was *taille*, in this story about a woman who pays her husband's debts with sexual favors: "And if so I be faille," she declares, "I am youre wyf, score it upon my taille."[2] That is, "put it on the tab"—although *tail* in fourteenth-century England apparently had the same slang meaning that it does today.

We only really learn about tallies in detail when governments adopted them. Shortly after 1066, Norman kings instructed local sheriffs to record tax assessments on tally sticks, break them, and present the "stock" to the Royal Exchequer, or treasury; payments for these also would be due twice a year, as in local village practice. But in this case too, tallies could become a form of circulating currencies. When kings ran short of cash (which they generally did), they would begin selling off some of their "stocks" before maturity from merchants who did have access to cash—at a discount, since the buyer could not be absolutely sure that the sheriff responsible for squeezing that amount of wealth out of the inhabitants of that particular district actually would be able to collect the required amount. Before long, a significant market developed in discounted tallies—basically, the equivalent of government bond markets—since English merchants preferred them over specie or kind when traveling through territories infested with thieves and bandits.

Another expedient for cash-strapped monarchs was to simply send out royal agents to appropriate things they needed from some hapless townsman or villager, record its value on a hazel twig, and leave the stock with the victim. There were numerous complaints about the practice in popular poetry, such as in "King Edward and the Shepherd":

I had catell, now have I none;
They take my beasts, and done them slone
And payeth but a stick of tree.[3]

These royal stocks could also be sold at a discount, but in this case the discount was enormous, since as one might imagine, collecting on debts owed by the sovereign was often extremely difficult.

Modern money systems only developed, in fact, once royal debts became so dependable that debt tokens could indeed be used as a dependable currency. China was already experimenting with paper money in the Middle Ages; the first forms of Chinese paper money, significantly enough, were ripped in half exactly like tallies. But the real breakthrough came with the invention of central banking systems like the Bank of England, created in 1694 when a consortium of London merchants made a substantial loan to King William, and in turn received from the monarch the right to lend the money he owed them to others in the form of paper notes. Within a century, almost all major governments adopted a similar system. Yet even as late as 1826, the Bank of England itself still kept its own internal records in the form of hazel wood tallies. These were destroyed—along most of the Parliament buildings—when under orders from the treasury they were put in the coal furnaces underneath the House of Lords and burned, setting off a chimney fire that rapidly spread out of control. Like love, then, or oaths, tallies were unique and crucial, and yet also so ephemeral.

11.5 English tally

Notes

1. See Angus Charles Graham, trans., *The Book of Lieh-tzu: A Classic of Tao* (New York: Columbia University Press, 1960), 179.

2. "The Shipman's Tale" is part of Chaucer's *Canterbury Tales*, which were written between 1387 and 1400. See Larry D. Benson., ed., *The Riverside Chaucer* (Boston: Houghton Mifflin, 1986).

3. See F. J. Snell, *Customs of Old England* (New York: Charles Scribner's Sons, 1911), 294.

12

SHARING

MARIA BEZAITIS

12.1 Scene from the early days
of the sharing economy

The photograph is of an espresso maker offered to the home-owner by someone who stayed as a guest in her home.[1] The host accepted this espresso maker as a gift from her guest, a gift that reveals a specific set of circumstances: the host had helped her guest with packages that needed to be shipped, offering her own home's address, in lieu of his own out-of-country address, as the destination. Consequently, the guest was able to avoid hefty shipping fees. The guest then purchased an espresso maker for the homeowner and in explanation told her, "I was sure your own would break someday."

This isn't just any guest and host, and perhaps it isn't just any espresso maker. The host is a homeowner who lists her home through Airbnb and offers a room in it to visitors for thirty-five dollars a night.[2] Airbnb is a digital platform launched in 2009 that enables short-term accommodation rentals. It markets itself as a "trusted community marketplace for unique spaces." This particular host's thirty-five dollars per night fee includes access to a bedroom designated for the guest, access to and use of a shared bathroom, and access to and use of the home's common rooms, including the kitchen and living area.

The kinds of reciprocal gestures on the part of host and guest described above make sense if we assume a context of familiar relations, where gifts are exchanged more routinely, and where values like obligation, generosity, and trust are embedded in the relations themselves. We tend to do things like lend our home's address to people who we know, people to whom we feel close. The population that might know anything about the appliances that reside in someone's kitchen is even smaller. Familiarity with the state of someone's espresso maker is not just a question of access to spaces. Familiarity is a question of access to the routines in which the espresso maker plays a role—everyday practices that produce the sense of belonging that differentiates the home from any other social place. The guest's remark—"I was sure your own would break someday"—underscores the guest's experience of belonging, a fluidity between what's mine and what's yours in the home, and suggests that he has participated in the routines in which that espresso maker figures prominently. One would only know of the fragility of the espresso maker through repeated interactions with it. Further, we might be inclined to assume that these occasions involve some kind of social contact between the host and guest. Our own experience underscores the fact that intimate knowledge of an everyday thing in someone's home does not emerge from interactions between strangers.

Homes are domestic spaces, and arguably they are the most important personal spaces that belong to an individual, couple, set of roommates, or family. Homes are places where people nest their personal artifacts, where they nurture their personal habits and routines with people they know or grow to know, and where they retreat and recover from a world of people they barely know. While homes are sites of longer-term commitments, they have always been places where the personal and economic have been inextricably, though often invisibly, woven together. Whether through the division of labor between income earners and domestic work, or the relationships between homeowners and service workers, the home has a long history of personal and economic entanglements.

Homes are emotionally dense resources, and all kinds of feelings matter to the value produced by Airbnb. Excitement, curiosity, and comfort are equally important emotions to experience for hosts and guests alike. The feeling that stands out, however, is that of belonging: the sense imparted to the guest that this home that belongs to someone else is also your home. "Belong Anywhere" commands Airbnb's home page, exhorting guests to take the listings personally and make the home reserved their own. Indeed, reinventing experiences of belonging may be the real point of Airbnb, and most materially for the hosts. While guests gain the right to belong where they don't by any historical measure, the Airbnb platform makes possible a new relationship of belonging between the individual and their home, and one that in so many cases absolutely requires the presence of strangers. "Hosting helps them make ends meet," says Airbnb of its hosts.

One might assert that our espresso maker is simply a form of payment, made in response to the host's gesture to make her home available as an address for packages shipped to the guest. Payment is always a kind of exchange, of one thing for another. I give this, and in exchange I get something of some equal value. Platforms including Airbnb have been identified as part of a market move toward "collaborative consumption" and "sharing economy"—phrases that emphasize the personal and social aspects of economic relations where payment is often not limited to a monetary transaction. Many Airbnb guests bring gifts not because they are asked to but rather because they decide to do so, independently of the host. Once in the home, these gifts— bottles of wine or boxes of chocolate—create opportunities for interaction between hosts and guests.

With Airbnb, we have tens of thousands of examples of everyday people turned into host-entrepreneurs, who capture and create value by making the very rooms of their homes available to strangers for an overnight stay, or a few days, or week. Rejuvenated in use by strangers, these rooms had a life before

Airbnb as home offices or playrooms, extra bedrooms for what are now empty nesters, extensions of the interests of homeowners. Airbnb produces what are new relations between people and people, and people and things, with hosts and guests engaging in something that looks more like short-term living together and formerly personal objects transformed into vital mediators between strangers. Indeed, Stanley Milgram's concept of the "familiar stranger" takes on more intensiveness and sustained presence in everyday life, as the setting for the familiar stranger shifts from the broad urban landscape Milgram discussed in his 1972 paper to platforms like Airbnb.[3] Today's familiar stranger is not the repeatedly observed stranger at a distance described by Milgram. Today's familiar strangers are the individuals who take up short-term residence in our home, the bedrooms those guests inhabit, or even the espresso maker offered as a gift between strangers.

An obvious question is whether or not Airbnb hosts provide a service that is similar to the hotel industry and different only in scale. A few key points suggest that we should not assume that leasing a room in someone's home is like staying in your favorite bed and breakfast or hotel. Three facets of these host and guest relations should be stressed here.[4]

1. *Some portion of Airbnb hosts, along with their families, live in the homes they make available to guests with no physical separation.* In many of these cases, hosts and their families share use of the rooms made available to the guest (bathroom, kitchen, and/or living area), with the exception of the guest's bedroom. Hosts and guests often negotiate access to shared rooms. These shared rooms are also important settings for social interaction.

2. *There is no in-person accounting that takes place between host and guest.* Airbnb runs the financial transaction through

its online reservation system so there is no moment of settling the bill between host and guest. Technically speaking, hosts are paid not by the guest. They are paid by Airbnb, which takes a broker's fee from the cost of the reservation. The payment clears in the host's bank account within some time frame of the guest's departure.

3. *Some of the hosts spend social time with their guests, and the relationship is not always bound by the time frame of the reservation or physical limits of the home.* In research conducted in Portland, Oregon, some hosts talked about inviting a particular set of guests out on a social outing with them or staying in e-mail contact with a guest. Others have joined their guests for dinner at local restaurants or a drink at the local bar. For some hosts, the process of getting to know their guests is integral to the value in making a room in their home available. This led at least one set of hosts in Portland to proclaim the following: "It's really nice to have people here for longer stays. They meld into the comings and goings of the day. It turns it into almost a family-like thing."

More recently, the sharing economy has become less known for its sharing and more widely recognized as a means to asset monetization—a frame that resets the interest on extracting value from the individual and away from the more amorphous kinds of value that can emerge between people. Over the last two years, Airbnb has focused more on the individual interests of hosts and guests, and brought a higher degree of expediency and predictability to the experience by introducing features like "instant booking" and hospitality standards that target higher, more consistent reservation rates. What felt like serendipitous encounters between hosts and guests in the early days of Airbnb are now proposed as explicit suggestions on the platform: "You invited your guest in … now consider inviting them out!"

The espresso maker is the figure that makes it possible for individuals to make certain kinds of exchanges in a context that doesn't yet provide an adequate justification for the exchange. Payment or not, the espresso maker is a protagonist in Airbnb stories about strangers together in that it advocates for a personal dimension in the context of a relation that lacks such a dimension but needs it. As gift, the espresso maker may be one of the most important actors in the sharing economy, lending familiarity to unchartered relations, providing crucial vehicles for the expression of affect between people. Its capacity to open up existing social ties is perhaps also a sign of more to come: more commercial value that seeks to produce short-term relating between people.

Indeed, the short-term-ness of the relation is precisely the point as well as what makes the espresso maker such an unsettled and unsettling thing. In a context of longer-term relations, the historical site of values such as reciprocity and obligation, this kind of exchange would be less surprising. Yet digital platforms like Airbnb are showing us that the gift is vibrant in its ability to contract to the short-term, and yet still carry the weight of values that once belonged to longer-term commitments and relations.

Notes

1. The photo was captured as part of an ethnographic research study of Airbnb hosts in Portland, Oregon, in 2013.

2. Airbnb is privately held and headquartered in San Francisco. According to its Web site, Airbnb is available in over 33,000 cities and 192 countries worldwide.

3. See Stanley Milgram, "The Familiar Stranger: An Aspect of Urban Anonymity," in *The Individual in a Social World: Essays and Experiments*, ed. Thomas Blass (London: Pinter & Martin Ltd, 2010 [originally published in 1977]).

4. These points emerge from the Portland study referenced above, and hence only speak to settings where common spaces are shared as well as where hosts live and are present in the homes that they open up to guests.

13.1 Reverse of four shilling colonial currency from Delaware Colony, signed by John McKinly, Thomas Collins, and Boaz Manlove, and printed by James Adams

13
LEAVES

WHITNEY ANNE TRETTIEN

In 1738, Benjamin Franklin began printing leaves of money. *Actual* leaves. By pressing thickly veined foliage into a cloth-covered plaster mold, Franklin had devised a method for casting relief blocks directly from nature. The resulting prints show cut sage, a trio of raspberry leaves, and delicate fern fronds, embedded in the coarse weave of a crooked cloth. To modern eyes, they are charmingly, startlingly realistic: one imagines the leaves being plucked from a growing plant, gently pressed into the cloth—a little more firmly on one side perhaps than the other, muddying the block's imprint. Their fossilized imperfections sate our own thirst for realism, seeming to connect us directly to *life* in an era before photographic reproduction. Printed on paper money, though, they warned eighteenth-century colonists that *"To Counterfeit is D E A T H."*

Nature printing—as old as the cave art at El Castillo, where humans first pressed their painted hands to stone—originally came to the colonies by way of Francis Daniel Pastorius. A Franconian transplant to Pennsylvania, Pastorius had brought with him the German craze for plant impressions, adorning the margins of his book with leaves of the New World. Pastorius may have taught his friend Joseph Breintnall, an avid botanist, how to print directly from nature, because by 1731, Breintnall was publishing beautiful specimen books of leaves for use in identifying plants. When Breintnall wrote an article on the rattlesnake herb for Franklin's 1737 issue of *Poor Richard's Almanack*, he probably suggested illustrating it with a leaf print. Although a direct impression was untenable for an almanac with a print run of around ten thousand, the challenge spurred Franklin to experiment with casting molds that could produce a relief block directly from organic material. He hit on a method that worked, and by 1738, began printing leaves on Pennsylvania's currency.

There's a simple reason why Franklin seized on leaf prints for paper money: they're difficult to counterfeit. England kept early America chronically short of gold and silver coin, thereby fiscally tethering the colonies to their mother country. By 1690, the scarcity of specie had pushed Massachusetts to disseminate its own payment objects in the form of government-issued banknotes. South Carolina followed in 1703, as did the remaining colonies shortly thereafter. Within two decades, America was awash in not only paper money but also its identical twin scourge: counterfeit bills. Easy to produce and difficult to detect, this specious currency undermined the public's trust in banknotes, threatening to destroy the delicate transactional network that was emerging within and between the colonies.

At first, what are known as "stub books" proved a useful check against fraud. Elaborately designed notes were printed or even hand drawn into a numbered notebook, and then cut away, leaving part of the design. By aligning a paper note with its bookbound stub, the bank could confirm its authenticity and track its possession. Of course, this system proved cumbersome at scale, and offered little guarantee to the common citizen accepting paper money in exchange for goods and services. So governments turned to the note's material substrate. When added to the paper, special watermarks, dyes, silk threads, and flecks of mica visually affirmed a note's value, while typographic extravagances further deterred the would-be counterfeiter. Franklin, for instance, spelled "Pennsylvania" differently on each denomination, making it possible to detect when someone had scraped away a lower denomination and replaced it with a higher one (a common form of counterfeiting). Later in the century, the

13.2 Michael Pexenfelder, *Apparatus Eruditionis tam Rerum quam Verborum per Omnes Artes et Scientias* (Nuremberg: Sumptibus Michaelis and Joh. Friderici Endterorum, 1670); in the margins of this copy, Francis Daniel Pastorius has made nature prints, the earliest-surviving American nature prints

Maryland printer William Green used multiple typefaces, sizes, out-of-place punctuation, and the occasional random letter to discourage what were by then British and Royalist counterfeiters flooding the market with fakes as a form of economic warfare against the revolting colonies. In the century of the Enlightenment, when printing was everywhere illuminating new ideas, colonial currency printers deployed their trade for deliberate obfuscation.

Perhaps the most clever anticounterfeiting technique was nature printing. Forging a leaf print is not merely difficult, as it is with Green's typographic oddities, but also impossible, since pressing delicate organic matter into a metal printing block destroys it. Only stealing the block from the printing house could ensure an exact reproduction of a leaf's intricately veined, perfectly unique structure. At the same time, the cyclic nature of plant life—its growth and decay, its incessant *becoming-something-else*—prohibits the forger from ever recovering that moment at which the leaf's image was frozen in plaster. Franklin's innovation, then, is that he shifts the burden of counterfeiting from copying the *content* of a note to discerning and iterating the *process* of its reproduction—even as, ingeniously, that very process prevents the thing, the leaf, from ever being reproduced in the same way again. By arresting time in space, leaf prints exploit nature's rare artistry for artificial ends, transforming the ostensibly irreproducible into widely circulated notes of exchange.

This tension between organic life and its inorganic reproduction plays into a broader contemporaneous debate about the relationship between the natural and the artificial. In the wake of the experimental and mechanistic philosophies of the late seventeenth century, nature increasingly seemed to operate like a machine—a sentiment confirmed by the tiny, ordered universes

13.3 Nature prints of leaves made by
Joseph Breintnall and Benjamin Franklin,
circa 1740

discovered under microscopes and on distant moons. If this
were so, then humans presumably could, with enough scientific
knowledge, exactly reproduce it—a project that the eighteenth
century pursued with enlightened gusto, designing ducks that
defecated, automata that wrote, and confectionary bouquets that
looked and smelled like real flowers. On the one hand, Franklin's leaf prints exemplify this mechanistic worldview, suggest-
ing that the humble printer might, with a clever enough method,
reproduce nature's artistry directly, not filtered through a human
engraver.

Anticipating William Henry Fox Talbot's *The Pencil of Nature* by a century, these early banknotes thus hint at the changing character of technological mediation on the cusp of electrified modernity. On the other hand, by using God's design to *deter* counterfeiting, Franklin implies the impossibility of humans ever truly imitating nature through artifice—or at least the potential sacrilege of such a task; for anyone who attempts to surpass "the Greatest and best Engraver in the Universe," as one nature print read, risks blasphemy. By choosing to print leaves, then, Franklin deploys both a material *and* cultural proscription against forgery, even as he playfully gestures at the printer's own godlike abilities. It's an irony that only succeeds because Franklin is printing on payment objects.

Franklin was an early proponent of paper currency, penning a short defense of its use in 1729. Responding to the widespread distrust of paper money—due to both inflation in Europe and rampant counterfeiting in the colonies—Franklin proposed securing a note's value against *land*; "for as Bills issued upon Money Security are Money," he writes, "so Bills issued upon Land, are in Effect *Coined Land*."[1] The value of silver and gold fluctuates, Franklin argues, as more precious metals are discovered. By contrast, the quantity of available land could only increase if a population decreases—which is highly unlikely to occur in the burgeoning colonies. Indeed, as immigration rises, the value of land in America would too, bringing up the value of the colonies' paper money with it. Although Franklin's reasoning has been widely discredited by modern economists, the treatise won him the profitable job of printing Pennsylvania's paper money, which he adorned with images of leaves, taken directly from nature. In addition to serving as anticounterfeiting technologies, Franklin's leaf prints pun on his own backing theory, demonstrating how immobile land might fiscally anchor freely circulating specie. Authority inheres not in the material substance of the paper itself but rather in the bountiful land that prints it.

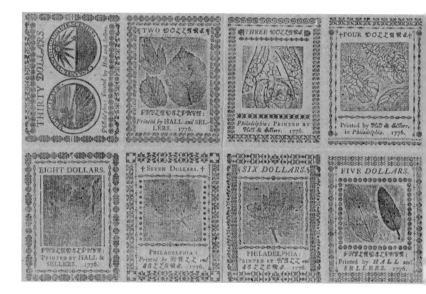

13.4 Uncut sheet of Continental currency with nature prints to discourage counterfeiting

Whether Franklin is shooting quips at mechanists or promoting his own economic theory, it is paper money's status as a transactional object that makes his punning possible. As an artifact, a paper bill is fraught with paradox. It is at once a singular thing, capable of being folded in one's pocket or exchanged for something else, and indistinguishable from all others in its class; its material form must express both uniqueness from *and* identity with all other bills of the same value. As such, the design of the bill must be rare enough in the wider world to thwart copying but similar enough within a series to instill confidence that any *particular* note will be worth the same as others in its denomination. While printing makes the mass reproduction of payment objects possible, this blessing is also paper money's curse, since a note that is easy to copy is also easy to counterfeit. By printing directly from nature, Franklin gathers together cultural proscriptions and beliefs in service of authenticating the strange materiality of paper money.

Note

1. See Benjamin Franklin, *A Modest Enquiry into the Nature and Necessity of a Paper-Currency* (Philadelphia: Historical Society of Pennsylvania, 1729), http://franklinpapers.org/franklin/framedVolumes.jsp?vol=1&page=139a.

Further Reading

Cave, Roderick. 2010. *Impressions of Nature: A History of Nature Printing.* London: British Library.

Green, James N., and Peter Stallybrass. 2006. *Benjamin Franklin: Writer and Printer.* New Castle, DE: Oak Knoll Press.

Levenson, Thomas. 2010. Benjamin Franklin's Greatest Invention. *American History* 45 (4): 26–33.

Musella, Francis. 2010. Benjamin Franklin's Nature Printing on Bank Notes. In *Money on Paper: Bank Notes and Related Graphic Arts*, by Alan M. Stahl. Princeton, NJ: Princeton University Library.

Scott, Kenneth. 1957. *Counterfeiting in America.* New York: Oxford University Press.

14

MINITEL

JULIEN MAILLAND

As I was growing up in Paris, there was a strange plastic cube sitting in the living room of my family's apartment. It was beige in color and would have nicely tied the room together in a French version of *That '70s Show*. With the press of a button, one of the cube's side panels would fall open to reveal a keyboard on the inside of the panel and tube screen lodged inside the cube. It was a computer. My parents had gone to the post office one day circa 1983, where the civil servant of the Post, Telegraph, and Telephone Ministry (PTT) handed them the machine, with instructions to plug it into the phone line to access a world of online services. Not too keen on new technologies, and not one to be told what to do by the state, my dad declined the offer. Refusal wasn't an option, the post office lady said; the paper phone book was being discontinued, and the only way to find phone numbers going forward would be to take the machine and connect it to one's home phone line. What's more, the machine was free. It was called a "Minitel."

Started in 1979 as a high modernist state program implemented through the state-run monopoly telephone service, Minitel became the world's first mass market for e-mail, online chat and bulletin boards, e-commerce, online games, and of course online porn. In 1989, while the World Wide Web had yet to be invented, "Minitel made France the world's most 'wired' country."[1] Everyone was online. In 1993, the network's peak year, French people collectively spent ninety million hours online. The almost seven million private Minitel terminals in a country of roughly fifty-five million people at the time, and shared terminals located in public spaces, connected to twenty-three thousand sites encompassing as broad a range of content and services as today's Web, multimedia applications excepted.

The Minitel machine was placed on the carpet of my parents' living room, by the phone jack, where it remained for the next thirty years. Carpets were changed, grade cards mailed in, birthdays, funerals, summer vacations at camp or with the family, laughs, tears, probably some champagne was spilled on it at one point or another, military physicals, college, law school, family reunions … by the time France shut down the network in 2012, the machine my parents had been handed three decades prior by the post office lady was still booting as fast as on its first day out of the Telic Alcatel factory, pride of the French telecommunications industrial renaissance. Initially built to last ten years, the machines never failed. They were as sturdy as a rock, as sturdy as the French state likes to see itself.

I wasn't allowed to touch the Minitel terminal, though, unless my parents needed to use it, in which case I'd be called on to play operator. Overall, the family unit rarely used it; it had the taste of forbidden fruit because every second spent on it cost outrageous amounts of money. Unlike with today's Internet, where a lot of content is still free, and where many for-profit sites still lack a clear and sensible business model, the Minitel business model was clear, simple, effective, and the same for all sites because although the sites were privately run, the network itself was statist.

Users were billed by the minute. There were multiple tiers of pricing depending on the services, from free or cheap for customer-support sites, to 1.41 euros per minute (the equivalent of 4 US dollars in 1983) for the most expensive professional databases and porn sites. Many a French kid remembers running into deep trouble after using the Minitel in their parents' absence, thinking they would get away with it, only to find out there was no

14.2 A Minitel in its natural
environment, December 2011

escaping the charges that would appear on the monthly phone
bill. Daniel Hannaby, founder of the popular *SM* chat site, recalls
that a woman sent his company a letter begging for mercy after
she racked up a 50,000 French francs bill, roughly US$24,000 of
the time. She claimed she had left her Minitel on by mistake.[2]
Michel Landaret, another pioneer of online chat rooms, recol-
lects that his best customer incurred a 225,000 French francs
bill (US$100,000 at the time) over the course of two months.[3]
The high cost of services and pay-per-minute model sometimes
created cultural patterns of use shared throughout society. All

French people old enough to have tinkered with the machine remember racing the clock to use the digital phone book: dialing, connecting, typing as fast as possible, ... and disconnecting before three minutes could run their course, ... then repeating the process as many times as necessary to find their contact's information. The first three minutes of that service were free, which explains the nationwide habit that now remains a shared antique in the collective national memory. To this day, #Minitel serves as the rallying cry for those sharing such early digital adventures; the Twittersphere is filled with nostalgic puns and sarcastic jokes day in, day out.

The presence of the Minitel in families' homes was a love-hate relationship. No matter how dangerously costly, controversial, or scandalous its existence, the terminal would always be there, in every household, a staple of state high modernism, the French cultural synonym of a picture of Joseph Stalin in a USSR kitchen. It seemed that you weren't a good citizen unless your Minitel, a sturdy, sealed, opaque plastic box as impenetrable as the state itself, was there, sitting on the ground, on a console under a tiled staircase, on an antique buffet—and by antique I don't mean a fake bought from Pier 1 Imports but rather a buffet passed from generation of farmer to generation of farmer until it makes its way to a Parisian apartment where it is forever stuck because they then build an elevator in the staircase of your nineteenth-century quarried-stone building and all of a sudden the staircase is too narrow to let the buffet go down, and so is the window, so the buffet gets trapped in the apartment, with the Minitel towering on it being the only way to tell what generation it is. Just like rural buffets were a staple of analog France, so became Minitel in digital France. In fact, its use was at times *required* by the state, such as to register for military service or with public universities, or to access certain other state services.

14.3 As part of its high modernist endeavor, the state introduced a new "ABCD" keyboard layout for the first 110,000 machines to roll out of the Telic Alcatel factory. Consumers preferred the traditional French "AZERTY" layout, which was reverted to in subsequent series

14.4 Open, closed

There was no more escaping it than there is escaping the tax collector.

But what made the Minitel system so successful that it became part of the national myth was first and foremost its billing system. In order to make the retail online system perennial, something no other country was able to achieve in the eighties, the state had to jump a few hurdles, not the least being to build critical mass. Being part of the information industry, the Minitel economy was subject to network effects: the network is only as good for any given member as its number of participants. In this case, the value of the network to each *user* depended on the

14.5 Minitel user guide, circa 1983

number of *services* available on the network. Conversely, the value of the network to each *service* provider depended on the number of *users* in the network. In order to solve the chicken-and-egg problem (which comes first, user or service?), the state forced the free terminals onto the people in order to create the large customer base that would make it worthwhile for service providers to create content: the more potential users, the greater the likelihood of a return on investment for creating and running a costly database. Combined with the free online phone book that the PTT Ministry maintained as a hook, the strategy proved effective in priming the pump.

What made Minitel so successful, however, wasn't just this astute move. It also wasn't just the beautiful simplicity of its interface, nor the rugged, nonintrusive design of its free hardware. One of the keys to the success of Minitel was the billing system embedded in the core of the network, called "kiosk." Under this system, the PTT Ministry, which operated the network, would collect fees from users of privately maintained online services as they connected to these services through the network's gateway dubbed "kiosk," and then rebate two-thirds of those fees to the content provider. The *kiosk* metaphor was used for several reasons. One of them was that the regional press industry, which understood early on that online services would shatter its business model, waged a war against Minitel, at times accusing it of being the spearhead of a new "big brother" state (video cameras connected to the central government were rumored to have been installed inside the terminal's screen so that the state could spy on citizens), and sometimes warning of the chaos that would result from an unregulated system, unless the press industry be given a monopoly to distribute content over Minitel. The reference to an analog kiosk—a centralized point where the newspaper retailer receives newspapers, distributes them to

14.6 The "Magis" Minitel terminal included
a credit card reader to make safe online purchases

the people, collects fees from readers, and then rebates a large
chunk of that fee to the newspapers—made it seem as though the
Minitel system had been tailored for the press. Along with some
other favors from the state, such as granting the press industry a
monopoly over access to the kiosk system, this move appeased
the lobby.[4]

The kiosk system was brilliant in several regards. Billing
enforcement was foolproof because the connection fees were
assessed on the phone bill, and failure to pay would result in the

14.7 A user gets on the popular 3615
ULLA chat room, December 2011

basic phone service being cut. There were no barriers to entry
because the Minitel functioned on a pay-per-play model with no
cover charge. In contrast to US online services in the eighties,
most of which failed because they did not reach critical mass
and were crippled by high barriers to entry, no monthly subscription was necessary on Minitel, nor was any prior registra-
tion, because the kiosk billing account was automatically tied to
the existing phone line. Minitel was as plug and play as it goes in
this respect. One could just try it, with no commitment whatsoever; nothing was simpler since the terminal had been provided
for free and was sitting in the living room to start with anyways.

Je voulais de la compagnie
j'ai tapé perroquet.
Pour avoir la paix
si je tapais
poissons rouges ?

ROCCO
C'est le
plus beau !

Télécarte 50   France Telecom

14.8 Minitel-themed pay
phone smart card, circa 1992

And what are a few francs to try out an intriguing brave new
online world?

An important part of this new world was playful and pink. A
staple of my life as a French teen wandering the streets of Paris
in the eighties was being surrounded by ads for online text-
based porn. For despite state claims to the contrary, online porn,
dubbed "pink Minitel" (Minitel Rose) by the French, was by far
the most successful of all online services, the one that *made*
Minitel, just like porn movies had made the VHS-tape system.
The pink Minitel is what made the overall network operation
financially sustainable for the state. In turn, the kiosk billing
system, with its foolproof privacy protections, is what brought
online porn to the streets, the home, and ... the workplace. Minitel was completely anonymized.

14.9 Minitel-themed pay phone smart card, circa 1980s

The kiosk system had already established the state as the inevitable center of all digital economic activity. Just like the centralized Colbertist road-network design had, three centuries before, "severed or weakened lateral cultural and economic ties by favoring hierarchical links," which forced all traffic through Paris, the kiosk embedded the state at the center of all things networked, and reproduced in the digital age the existing French state-society organizational model.[5] The fact that all Minitel traffic had to transit through the centralized state's computer nodes controlled by the telephone operator also had a major advantage in terms of privacy protection. The network resembled an hourglass, in the middle of which the state served as a nontransparent buffer. It anonymized upward connections by masking the identity of the end users for the content providers. On the downstream side, the state anonymized the source of the communications that had taken place. That is, when the phone bill would show up in the mail, it would indicate that the user owed the state X amount of money for Y minutes of connection to "Minitel services." The actual sites accessed were never individually named. This system ensured peace in family units. Children would still get punished for unauthorized connection time, but parents would not have the nature of their Minitel activity disclosed to each other through the phone bill. The privacy system embedded in the billing system also made porn popular in the workplace. White-collar workers used Minitel to order goods and services for their company, or check corporate records and the financial health of potential business counterparties. Sneaking in a bit of sex chat in the middle of the workday became commonplace, and there was nothing that human resources and compliance officers could do about it. Digitally monitoring employees behind their backs was illegal. Since the ministry anonymized the source of the downstream flow on the monthly corporate phone bills, lest employers put a human monitor behind the shoulder of every employee, there was little they could do.

When the Minitel network was shut down by France Telecom on June 30, 2012, there were hacker parties and "wakes" throughout France to mourn the system. But its legacy lives on in the Apple ecosystem. The design of the 1984 Apple Macintosh resembles (perhaps too closely for many French observers) the design of Minitel.[6] The Apple ecosystem, just like Minitel, is censored, in the sense that third-party providers can only push their content through the network after receiving Apple's prior approval. Apple's app billing system is also mirrors the kiosk: when buying an app, the user pays the network operator (Apple) for the content pushed by a third-party provider. The network operator then rebates the content provider a part of that fee. The rebate level happens to be similar to that applied by the French PTT: the content provider receives two-thirds of the fee paid by the user, and the network operator keeps one-third. The kiosk lives on, though the pink iPad has yet to be invented.

Visit the Minitel museum at www.minitel.us. We won't bill you by the minute. It's free.

Notes

1. James Gillies and Robert Cailliau, *How the Web Was Born* (New York: Oxford University Press, 2000), 111.

2. Daniel Hannaby, interview with author, San Francisco, California, March 9, 2012.

3. Michel Landaret, Roundtable, "Minitel: La dernière séance 'Rendez-vous à jamais,'" *La Cantine*, June 29, 2012.

4. Jean-Paul Maury, interview with author, Paris, December 2011.

5. James Scott, *Seeing Like a State* (New Haven, CT: Yale University Press, 1998), 76.

6. Marie Carpenter, *La Bataille des Télécoms: Vers une France Numérique* (Paris: Economica, 2011), 561.

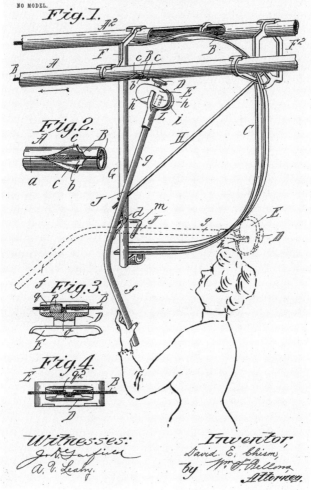

No. 754,424.

PATENTED MAR. 15, 1904.

D. E. CHISM.
CASH CARRIER APPARATUS.
APPLICATION FILED OCT. 3, 1903.

NO MODEL.

Witnesses:
J. L. Garfield
A. V. Leahy.

Inventor,
David. E. Chism,
by Wm. F. Bellew
Attorney.

15.1 Patent for cash carrier apparatus, 1904

15
RECEIPTS

JANE I. GUYER

Proofs of Purchase, Units of Account

Perhaps due to a childhood routine from the 1950s in Britain, I pause over the receipt itself that records the individual transaction and its physical fate: what it is (a piece of paper, a mark on a card, etc.); how it moves; whether and where it is kept; and what happens next. A receipt is a proof of an event in the past. It has the name of the issuing party, date of the transaction, amount of money transacted, and in the case of the purchase of objects, details of the material thing that changed hands. Some receipts also bear a signature that can verify the identity of the person who agreed to it as well as the titular organization for whom they acted. And it has a status in law and administration. We submit receipts for the year-end tax accounting to prove the kind of transaction that each one was: a donation, purchase, investment, and so on. In these days of digital record, we are still asked whether we would like a receipt for a purchase, however small. If we say "yes" and accept it, it probably ends up in the trash, even though—at the time—there seemed to be something worthwhile in holding it in one's possession, even for a few minutes. Perhaps we retain an aura of the conviction from other transactions and the past that receipts were crucial evidence that served to prove that event along with one's worth as a transacting person, even in the jural-legal sense. My status, veracity, and trustworthiness are backed by these physical proofs.

Receipts as well as other markers of transactions have to be curated if they are to acquire this social power in the first place and then maintain it across time. The element of ceremonial display and careful preservation is present, even if in mundane ways. As for dramatic ways, I have been wondering whether the oral trading cultures of West Africa in the past had poetic mnemonics for transactions, such as stories that might have been elicited by the figurative gold weights of the Ashanti system. The gold weights might be technologies not only for the actual weighing of the gold dust, for which purpose they seem inscrutably inexact, but also for creating a conversation between the parties to a transaction that slots its terms into both of their memories, in the same terms. My questions about gold weights and their characteristic modes of negotiation remain empirically unconfirmed so far. My effort to imagine them in use, however, suggests that we also have in our own transactional histories of practice a sense that the "unit of account" function of money is not entirely fulfilled by numbers and ledgers, and now digital archives; that there is more to the securing of transactional histories than the abstractions of account registers; and that there are still people, imagination and social relationships in the picture, with a certain ceremonial aura, whether in the technology or standard verbal exchange with which even minor transactions are sealed.

Receipts and Their Containers
The Cash Railway, Twentieth Century

My growing interest in units of account may well be given extra momentum, beyond the straightforward intellectual relevance of this topic to anthropological studies of money, by several instances where receipts have stayed in my mind—from life and from fieldwork. As a child, I ran errands to the grocery store. In those days my family was a member of a cooperative society, so we got dividends each year. Every purchase was an act of membership, so we children had to know our number. I

15.2 Gipe wireline cash carrier system
at Trowbridge Museum, Somerset

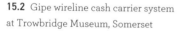

still remember it: 79041. The server behind the counter took our money, wrote up the transaction, put the whole package into a little container that looked like a funicular railway carriage on a wire, pulled a wooden handle, and whizzed it across the store to the accountant sitting in a little cubicle, behind a glass partition. A bell rang when it reached its destination at the till. The cashier unpacked it, put the change and receipt into the little container, and whizzed it back to the counter. I took the change and receipt back home, where my parents kept it to make the eventual cumulative tally against the co-op record.

The whole chain of the transaction had the aura of a mundane ritual: striking to watch, visible to everyone present, a demonstration of efficiency and transparency. That little machine was quite beautiful in its way: shaped like a lozenge, transparent in the middle section (to make everything visible), and with brass ends made shiny by the constant handling. The wooden pulley-handle was shaped like other handles, so it was crafted. Another version had the money and receipt put into a container that screwed into the railway system like a bottle into its cap. Eventually some stores had a kind of vacuum tunnel system (pneumatic tube) for these money and receipt containers. I suppose these gave the impression of automation and scientific advance. Sources about "cash carriages" or "cash railways" suggest that the first kind was a "cash ball," patented in the United States in 1881. The money-receipt container was a ball that just rolled down parallel lanes, like a miniature bowling alley. One lane sloped down from the counter to the till, for the money and bill, and the other sloped down from the till to the counter, for the change and receipt. The confidence of each party in the other was materialized at each stage, and allowed proof to be brought forward in a process that was either literally transparent in the glass components of the technology or clearly visible in its overall structure—to both parties to the transaction and everyone else in the store. The cash railway cars were whizzing back and forth all the time, from different counters in the grocery, each selling their own items.

I can remember a certain pride in being part of this whole process, beyond the routine childhood duty of "doing the shopping." At the time, there were many other stores where the counter person ran the till, so there were no such mechanisms, and no receipts either. Here you were a regular customer in a personalized transaction with a particular shopkeeper, not with an organization such as a cooperative, so both parties would just remember the purchase. Hence, I suppose, the embedded custom of exchanging conversation about something relevant to the purchase and being a "regular customer" of particular sellers, on regular rhythms of shopping for necessities.

Responding to my request for any information about these transaction technologies, a university friend from the 1960s, in England, wrote to say that cash railways now benefit from an enthusiastic following. The Cash Railway Website maintained by Andrew Buxton, contains many photographs of various designs.[1] Some of them are far more ornate than those I recall, perhaps giving emphasis to my suggestion that there remains a ceremonial element in monetary transactions that has to do with the memory and accounting aspects of money.

Curating Receipts, Twentieth Century

A small businessman's method: when I moved to the United States in 1965, I came to know a little about the business of a senior member of my new family who was a small-scale independent salesman making a lot of personal transactions all the time. He had a system of receipts that he kept in Havana cigar boxes, made from light wood and still emitting a slightly perfumed aroma, somehow evocative of moments of luxury. The boxes could be stacked neatly along shelves and preserved the paper from the inevitable dangers of damp in the only place he had for a home-based office: the basement. Another family member still keeps receipts in a cigar box in his study, with fifteen

15.3 Receipt box once owned by Bernard
Guyer's father, Sydney B. Guyer

cents written on the inside lid (see photo). I think it may be an
heirloom, and it still exudes that aroma. The memory function is
thereby enhanced in several directions. Every receipt of importance that cannot be filed in a filing cabinet is in that box. I imagine that many people have such boxes, associated as the "box" is
with "treasure," and thus with values that go beyond the material
object.

A colonial subject's method: when I was researching the popular monetary history of Southern Cameroon in the 1980s, I interviewed elders about the monies of the past. At one stage I was asking about colonial taxes, and an elderly man brought out a receipt and membership card for his contribution to the compulsory *société de prévoyance*. The provident society was a French colonial institution that accrued functions over the years. The fund provided some support in times of economic crisis and could supply capital for certain new activities that were in line with policy. Membership was compulsory. By the 1950s and the era of anticolonial struggle, it had become a kind of political surveillance tool (see Guyer and Mann 1999). Thirty years later, long after the end of colonial rule and abolition of the provident society system, this man could produce proof of payment from a secret place in his modest house in a rural village. There had been punishments for failure to belong to the provident society, so maybe he still thought he needed the receipt and record of payment in some way or for some purpose that he no longer understood. And yet he was not, apparently, willing to risk anything by failing to keep his receipt. This moment also reminded me of a scene from Mongo Beti's (1957) novel *Mission to Kala*, where one of the protagonists was carefully keeping his colonial tax receipt, which placed him in a specific "category" beyond the basic head tax, of which he was quite proud. It was a sign of his status.

Ironic Deployment, Twenty-First Century

Is it an accident or just a joke that the dust jacket of David Graeber's (2011) book *Debt: The First Five Thousand Years* displays a receipt—in this case for 00: nothing? Of course not. Once purchased from the publisher (I suppose with money), we are invited to take the contents of the book without incurring any debt at all, and are reminded of that debt cancellation every time we take the book off the shelf and open it.

Some Inferences, in the Experiential Register

Receipts are not just convenient artifacts of the "real" business, which is accounting. They are condensed forms for the varied affects associated with transactions: pride in one's own capacity as a transactor and concern about being challenged in the future. They relate the payer and payee forever. Receipts are props to confidence, proofs of one's own expertise, defenses against the dangers of accusation, and records of an event in time between at least two parties that performed some sort of agreement or acquiescence. And hence, perhaps, the aesthetics of the containers: modern mechanics and transparency for the co-op, borrowing from the technical prestige of the railways; the hint of luxury for the small businessman; the hidden places for the apprehensive colonial African subject; and the cover of a vastly popular, critical book about debt to make the ironic and social connection between author and reader.

My final observation on this personal roster of memories about receipts would be that the technologies of receipt management are clearly topics of conversation, which this short chapter extends further into the academic world. By provoking imagination and commentary, these receipt-conversations—probably like the Ashanti gold weights—restore the person, the experiential, the symbolic aura, and the political dimension to what we pretend to be impersonal transactions, reflecting the impersonal forces of the market and the bureaucratic state, and mediated by the universal equivalent we call money, which is described by number only.

Note

1. See The Cash Railway Website, http://www.cashrailway.co.uk.

References

Beti, Mongo. 1957. *Mission Terminée*. London: Heinemann. Translated by Peter Green as *Mission to Kala*.

Graeber, David. 2011. *Debt: The First Five Thousand Years*. Brooklyn, NY: Melville House.

Guyer, Jane I., and Gregory Mann. 1999. Imposing a Guide on the Indigene: The Fifty Year Experience of the Sociétés de Prévoyance in French West Equatorial Africa. In *Currency, Credit, and Culture: African Financial Institutions in Historical Perspective*, edited by Endre Stiansen and Jane I. Guyer, 118–145. Uppsala, Sweden: Nordic African Institute Publications.

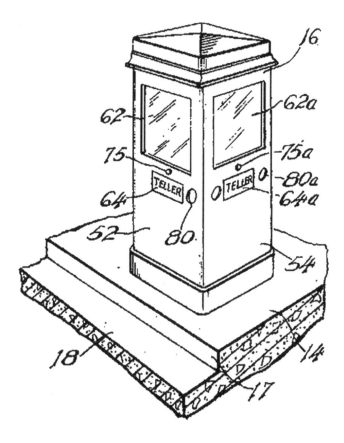

16.1 Early concept of a cash machine, 1955

16
ATMs

BERNARDO BÁTIZ-LAZO

Automated teller machines (ATMs) are ubiquitous, and most adults living in urban areas around the world have interacted with one. They are often portrayed in the media as the omnipresent point of contact with today's otherwise-digital bank. Lyrics in popular music from artists as diverse as Britney Spears and Radiohead have referenced ATMs. In the United Kingdom, poems about ATMs have been part of National Poetry Month celebrations. In spite of the ATM's cultural ubiquity, few stop to reflect on how the backbone of today's retail payments ecosystem came about, while others name it the harbinger of the death of cash.

Before ATMs, banks on both sides of the Atlantic had been trying to find alternatives to the brick-and-mortar branch since at least the interwar period. Automatic cash dispensing also predates widespread payments with credit and debit cards. The cash machine revolutionized banks' service provision by allowing after-hours access to bank deposits at a time when most people around the world worked in a cash economy and personal checks were outside the United States largely the remit of the well off.

Cash-dispensing machines resulted from a long chain of innovations rooted in the implementation of self-service in the 1950s, as evidenced in gas stations, supermarkets, ticketing for public transportation, and candy dispensers (see also chapter 18, this volume). A first device seems to have been deployed in Japan in the mid-1960s, but little is known about it. Instead, the most successful early deployments took place in Europe, where bankers responded to increasing unionization and rising labor costs by soliciting engineers to develop a solution for after-hours cash distribution. This resulted in three independent efforts, all of which saw the light of day in 1967: one in Sweden (Asea-Metior and the Swedish savings banks' Bankomat), and two in Britain (De La Rue and Barclays' Barclaycash; Chubb and Westminster Bank's Chubb MD2).

There was no single inventor but rather active collaboration between various groups of bankers and engineers to solve significant challenges. For instance, never before had electronic equipment been put to the test of the weather. Yet these early devices could all function outside, exposed to the elements; they were all stand-alone, clunky, unfriendly, and inflexible machines. They could do only one thing: dispense cash when activated by a token, such as a punch card (figure 16.2). Today in 2016, we typically use a plastic card containing a magnetic stripe or a chip, while in some countries they have begun to active ATMs with mobile phones. Early devices were highly unreliable, however, and there were no standards. For example, some banks' machines would keep the token used to active it, and the bank would return it to the customer by post once the account was debited.

By 1971, other manufacturers operated in Britain (Speytec-Burroughs), the United States (Docutel and Diebold), and Japan (Omron Tateishi), while devices had been deployed across Europe (e.g., Switzerland, Spain, France, Germany, and Denmark), Canada, and Israel. Following this period of relatively rapid deployment, an important judicial ruling in the United States found that Chubb, a UK manufacturer of bank equipment, and Diebold, a US competitor, had engaged in anticompetitive practices by colluding not to set up machines in each other's countries (*United States v. Diebold, Inc.* and *Chubb & Son, Ltd., Cr. No. 76–9A [N.D. Ohio 1976]*). A second case in 1981 gave Chubb the overall preeminence of its patents on the codification of the personal identification number (PIN). And in a third case,

16.2 Activation token for the Chubb MD2
cash machine, 1967

the US Supreme Court decided in 1984 that the ATM was not
subject to regulation limiting geographic growth.

Nevertheless, at this point in time there were two major limitations to the future prospects of this device. The first was that
cash machines would only work within proprietary networks—
that is, tokens of the local bank would only activate a machine
for that bank, and in some cases, only that bank location. This

was to change with the advent of encryption keys and message authentication codes, (MACs) which ensured that messages had not been tampered with while in transit between the ATM and financial network. These innovations then enabled shared networks accessible to everyone.

The second most important success factor was the advent of online authorization. Security was a paramount concern of bankers, who had to make sure that the right person debited the account at the point of the transaction. In the early years, though, cash machines operated off-line as a stand-alone device, or by transmitting information once a day through dedicated telephone lines or links to minicomputers in large branches. Hackers soon exploited the shortcomings of off-line operations. In Sweden over the Easter weekend of 1968, someone went up and down the country using the same (stolen) card in different machines, and this was not discovered until the holiday was over. In the mid- 1980s, people in the United States used Commodore C64 personal computers to manipulate the magnetic stripe of banking cards, driving from ATM to ATM to withdraw money. Online authorization saved the ATM business model and gave rise to a fault-tolerant computer system that still dominates the card payments market today.

It is worth emphasizing that the development of online communication with a bank's central computer became the overriding concern early on. The work of IBM has been greatly unappreciated here. IBM started with the first online trials of communication networks tied to a central computer in Sweden in 1968, and then deployed an online device in the United Kingdom for Lloyds Bank in 1972. For most of the 1970s, IBM engineers worked hard to develop the standards on which other elements of the payments ecosystem such as credit cards and point-of-sale terminals would eventually depend as well as what payments industry professionals term the "rails" and "pipes"—the infrastructures of electronic value transfer.

In the early 1980s, pioneers such as Chubb, De La Rue, Docutel, and Asea-Metior left the cash-dispensing business when they failed to keep up with these developments in computing and electronics. Those that were successful, like Burroughs, never achieved critical mass. Citibank abandoned possibilities to commercialize devices of its own design (called CAT-1 and CAT-2) for use on other networks, and instead continued deploying them throughout its own global proprietary network until the 1990s.

IBM had the marketing muscle, engineering expertise, and business contacts to take over the market. But in a strange twist of fate, the top brass at "Big Blue" decided to deploy a new machine, the IBM 4731, and the related IBM 4736 series of cash machines. These were incompatible with previous models, particularly the otherwise-successful and widely deployed IBM 3624 cash machine. Many banks evaluated the new IBM and refused to buy it because, in a stroke, IBM would make obsolete significant capital investments that the banks had already made. Worldwide sales of the new machines were low while banks began to purchase from other vendors. Consequently, IBM decided against further investments in payment technology, abandoning the potential to capitalize on its research and development as well as subsequent patents in chip-based activation tokens (now popular in many countries).

Around this time, two Ohio-based companies, NCR and Diebold, were working on technology that would enable them to effectively dominate the supply of ATMs for the next two decades. Both had their origins in cash-handling and storage equipment (NCR had been named National Cash Register until the 1990s; Diebold had been a maker of safes and vaults for banks). As result of the IBM 4731 and 4736 fiasco, NCR built software that emulated the IBM 3624 on its own devices and grew to become a leading world supplier of ATMs. Meanwhile, in 1984, IBM and Diebold formed a joint venture called InterBold that aimed at leveraging Diebold's self-service technology with IBM's global distribution system. Seven years later and in spite of growing sales, the joint venture ended as Diebold had not achieved the breakthrough it hoped for in the international market, while for IBM the joint venture failed to materialize expected returns partly as developments elsewhere had invalidated its strategy to link ATMs to expensive mainframes.

Capitalizing on deregulation and banks' diversification across retail markets, NCR and Diebold were instrumental in turning the cash dispenser "dinosaur" into today's sleek, multifunction ATM. The companies' innovations included new, customer-friendly video display units; programmable buttons alongside the screen; a shift toward dispensing cash horizontally, which reduced jams; and a growing menu of options, including money transfers and balance inquiries.

But NCR and Diebold were not alone. Consumers had overcome their initial reluctance and were actively engaging with this technology. Growth in the number of banks deploying ATMs across the world brought about an increase in the number of manufactures. These included big names such as Honeywell in the United States; Phillips, Olivetti, and Siemens Nixdorf in Europe; and others based in Asia such as Fujitsu, GRG, Hyosung, and Hitachi. Large European banks developed separate, proprietary networks even as US banks found shared networks (and their associated interconnection fees) attractive. Indeed, at one point there were some two hundred different networks in the United States, all servicing an increasingly mobile population willing to pay to access their money.

At a local level, advances in the functionality of the ATM freed staff at retail banking branches to engage customers in higher-value services, such as insurance, mortgages, and stock trading. ATMs thus proved to banks that there were alternative distribution channels to brick-and-mortar branches. They became the backbone of the payments system, and opened the door to

16.3 Sir Richard Summers inaugurating
a cash machine in Manchester, circa 1968

telephone and Internet banking. In this sense, ATMs enabled the explosive growth of retail finance during the last decades of the twentieth century.

At the same time, however, ATMs remained a significant capital investment, costly to operate. In spite of innovations with modular manufacturing along with its associated reduction in service costs and improved reliability (as the machine could continue to work when facing a small mechanical failure), the physical siting of ATMs remained expensive. The use of dedicated telephone lines limited them largely to bank branches, or high-volume nonbank locations such as busy train stations and big airports. This was to change with the advent of digital telephony and introduction of Microsoft Windows computer operating system. With Windows, the ATM could in effect become a terminal of the bank's central computer, enabling key operating functions such as remote diagnostics as well as the integration with credit card clearing networks, and with the side effect that there was no longer a need for a dedicated, bank-specific activation token. Through the credit card networks, one could use virtually any card at virtually any ATM.

In the mid-1990s, Mississippi-based Triton and Texas-based Tidel introduced lighter, cost-effective machines. These were cheaper to buy and run, as their characteristic advantage was that they contacted the bank computer only when activated by the transaction (as opposed to being permanently connected, as was the case with dedicated telephone lines). Triton and Tidel were instrumental in the creation of a whole new kind of entity: nonbank, independent providers also known as Independent ATM Deployers (IADs), which install and maintain machines located in places like grocery stores, commuter railway stations, universities, and even casinos.

The advent of IADs and integration of local ATM networks with the international Visa and MasterCard networks caused a huge expansion in ATM deployment in the late 1990s. The increasing popularity of debit cards also fueled growth. As of 2016, ATMs are moving into China, and to a lesser extent Africa, eastern Europe, and the Gulf states (with China seeing two-thirds of the global ATM growth in 2011). Not surprisingly, growth in the number of vendors, deployers, merchants, and suppliers of ancillary services across the world led to the creation of a "super niche" trade association in 1997 called the ATM Industry Association, which now hosts annual conferences in the United States and Europe.

From humble and uncertain beginnings forty-five years ago, the ATM and its associated technology are now ever present in our daily life. Today we are all asked for PINs not just in a banking context; many library book checkout systems require them, as do online marketplaces and self-service checkouts in supermarkets, illustrating how a solution for an industry-specific device has now spread across everyday life and around the world. The integration of ATM networks means we can travel almost anywhere with just a plastic card in our wallet, confident we will have access to our balance in local currency in places as far afield as the malls of Hong Kong, pyramids of Giza, or streets of Paris. One often finds that the user interface is pretty similar across the world, and we are even given the opportunity to choose our preferred language. But no matter where you find it, the ATM's core function remains the same: dispensing cash in thirty seconds or less.

Finally, we could ask whether the advent of the cashless, digital economy also heralds the end of the ATM. But the trend to abandon banknotes is not universal. As of 2016, the demand for cash continues to rise in most wealthy countries such as

16.4 Money-like ATM tokens
(ATM testing notes), circa 1985

the United States, the United Kingdom, and Canada, and the euro-area experiences a year-on-year growth of between 5 and 10 percent while ATMs are chiefly responsible for distributing as much as 85 percent of the banknotes in circulation in some of these countries. A recent *Economist* report cited Denmark and Sweden as nearly cash-free "outliers" (with Australia running closely behind). The use of cash in retail transactions is as evident as much as the growth in the use of alternative ways to make payments. As retail transactions move to the digital space to give bank customers greater choice, access, and freedom in the way they bank, however, so will malware increasingly focus on personal computers, tablets, and mobiles. In this context, ATM networks are bank assets that have been in place for some time, but more important, they remain within the banks' control. Indeed, for decades now the ATM has proven to be a trustworthy, reliable, and conveniently located channel for consumers all around the world. This is a great advantage to the ATM when compared with the host of alternative devices from which consumers can choose to interact with their bank. The future, then, might see the ATM not only remaining prevalent but also incorporating other self-service transactions (such as depositing cash or checks), and poised to become critical to delivering retail financial services.

Bernardo Bátiz-Lazo

16.5 (facing page) Details of GAYTM fascia (left) and Anzabella Darling (right), 2014

16.6 Queen Elizabeth II inspecting a Lloyds Cashpoint at the IBM site in Havant, UK, 1974

17

GREYBACKS

KEITH HART

17.1 Detail of Confederate $1,000 bill, featuring
John C. Calhoun

I left Cambridge, England, in 1965, to begin anthropological fieldwork in Ghana for my doctoral dissertation. The first thing I did when I arrived there was to head to the university to meet a famous American professor. He was a friend of my supervisor. I expected him to be interested in my research and perhaps even me, but all he wanted to know was whether I had hard currency, especially dollars and pounds. He explained that the official exchange rate was 1:1, a Ghanaian pound for one pound sterling, but he could get me an exchange rate of 1.5:1. I left it vague if I had any pounds and asked around town. To my horror, I discovered that the unofficial exchange rate was 3:1. This was the first of several culture shocks in those early days, usually involving economic transactions. I couldn't believe that a senior academic would try to cheat a graduate student, newly arrived in a foreign country and just beginning his field research. It didn't take me long, however, to realize that I was the cultural dope in this story. I actually believed that academics were an aristocracy of intellects who valued money less than ideas. This was not to be the last time I stumbled into a gray area where my assumptions about money and morality were challenged.

Eventually, I set up business with my landlord, Ananga, in a sprawling slum called Nima. He had the contacts and experience, and I had the money (and got the field notes). I soon discovered that I brought valuable expertise to money changing. I made regular trips to the University of Ghana to check the exchange rates in the English newspapers (there were no online searches at that time). For example, I knew the difference between hard and soft currencies, and the petty thieves who brought Ananga and me their pickings didn't understand that some moneys served as foreign exchange while others did not.

This led to a tense moment just after the military coup that displaced President Kwame Nkrumah. Around midnight, an army truck turned up with a number of heavily armed and excited soldiers in it. They had found many boxes of Egyptian piastres in the bedroom of Nkrumah's wife, Madam Fathia. Knowing our reputation as money changers, they wanted us to buy them. I pointed out that she could spend the money when visiting Egypt, but it had no exchange value outside that country (like the Ghanaian pound). What do you say to a disappointed soldier with a Kalashnikov? It was quite a relief when they left.

These are the high or low lights, of course. A lot of the business was mundane. But one evening, Ananga was over the moon. "We're rich," he shouted. He had just bought US$1,650 for a bit more than £50 (less than a tenth of the official exchange rate). At first I thought it must be counterfeit and complained that he should have waited for me. On closer inspection, I discovered that the dollars were in four notes: $1,000 plus $500 plus $100 plus $50, one each. Odd. It took me a while to notice the important part. The Confederate States of America had issued the banknotes in the course of the US Civil War; they were not greenbacks. They were greybacks.

I didn't know if they were ever legal tender after the Civil War (in fact, they weren't), but I imagined that in 1967, we were more than a hundred years after their issue. I gave Ananga his £50—he was used to the ups and down of informal commerce—and kept the Confederate currency as a memento of my fieldwork. They are still somewhere in a box of field notes at home. There is quite a market for them today, on eBay and elsewhere

17.2 Confederate $1,000 bill, featuring
John C. Calhoun and Andrew Jackson

17.3 Confederate $500 bill, featuring Ceres,
cattle, and a train crossing

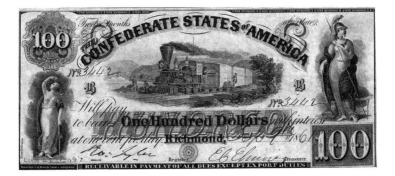

17.4 Confederate $100 bill, featuring allegories of justice, Minerva, and steam engine

17.5 Confederate $50 bill, featuring enslaved people working in a cotton field

(figure 17.5). They may well end up being an investment. One interesting aspect of this 1861–1864 issue is that great efforts were taken to make each note highly particular, and this means that some have the value of rare books.

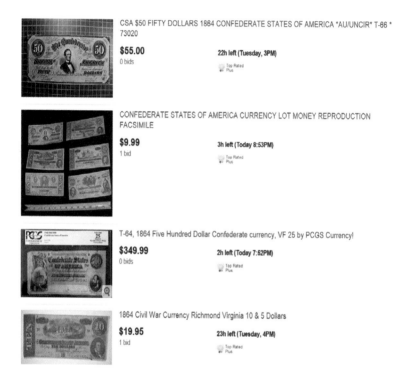

CSA $50 FIFTY DOLLARS 1864 CONFEDERATE STATES OF AMERICA *AU/UNCIR* T-66 * 73020

$55.00
0 bids

22h left (Tuesday, 3PM)

Top Rated Plus

CONFEDERATE STATES OF AMERICA CURRENCY LOT MONEY REPRODUCTION FACSIMILE

$9.99
1 bid

3h left (Today 8:53PM)

Top Rated Plus

T-64, 1864 Five Hundred Dollar Confederate currency, VF 25 by PCGS Currency!

$349.99
0 bids

2h left (Today 7:52PM)

Top Rated Plus

1864 Civil War Currency Richmond Virginia 10 & 5 Dollars

$19.95
1 bid

23h left (Tuesday, 4PM)

Top Rated Plus

17.6 Confederate bills of variable
price and provenance, eBay, 2015

It was inevitable, I suppose, that this trade in currencies leaked into money lending. And here we enter an even greyer area—morally, ethically, and economically. But it was quite an education, and I can't imagine learning what I did then had I not participated in it. I made loans to members of my familiar circle. For example, Atinga, one of my associates, often borrowed from me to sustain his impoverished gin bar business.[1] His problem was that his customers had no money, and so he more or less had to give the stuff away. One day, I found that he had made a loan using money I had just lent him. I was pretty angry, but he pointed out that people frequently came to him for repayment of loans and he could now send them after his debtor.

Apart from this intimate borrowing, I decided to lend money formally to relative strangers—so that I could learn something about money lending. This was an eye-opener. I soon realized that the only way I could find out about this ancient craft was by doing it myself. We usually read about the headline interest rates in informal finance—and they do appear to be astronomical, like 50 percent a month. Yet the key factor is the default rate, not the interest rate. And no moneylender would ever tell a researcher about their default rate, since they depend on the mystique that they do not allow default. I had a friend, a beer brewer, who was thought to be the richest Frafra woman, but she sold most of the beer on credit and sometimes could not find the cash to start a new brewing cycle. What she was rich in was young, male clients who owed her money and could be put to various casual tasks, such as heavy lifting and sharing her bed.

Money lending was only a part of our business. Receiving stolen goods was more important. But I became embarrassed by how much I was making. I extended my fieldwork by eighteen months solely on the revenues of my business with Ananga. I felt compelled to spend my wealth. It relieved my ethical

17.7 The author, a few years after the events of this chapter, playing Owari, a bead game, in Nima

quandary associated with those gray zones (and there were no ethical review boards to which one could go for formal or informal advice on the matter). It was unthinkable to make money from poor people. So I hired seven research assistants at one time, gave large rice, beer, and sheep parties, and donated sandals or blankets to old people.

Ironically this did not reduce my wealth: I was now seen as a big man redistributing to the people and became even more popular with the thieves. At last I succeeded in leaving Ghana without a personal profit and concealed how I got some of my richest field notes, the stories of money that began with my encounter with that senior academic on my first day in the field. Those Confederate greybacks told one story during the US Civil War, led Ananga to hope for another tale, and surprised me when I received them a hundred years after their issue. Those Egyptian piastres contained soldiers' hopes and sparked a young anthropologist's fear. My money changing and money lending stitched me into a community of confidants, friends, and thieves. Money is a means of communication. I don't know how I would have managed in Accra without it.

Note

1. For the case study, see Keith Hart, "Informal Income Opportunities and Urban Employment in Ghana," *Journal of Modern African Studies* 11, no. 3 (1973): 61–89.

18.1 "YOU ARE THE CASHIER"

18

THE SWIPE

MICHAEL PALM

The transitive verb *swipe* traditionally describes acts of theft and violence. To swipe something means to steal it, while taking a swipe at someone involves a punch or slap. Recently these actions have been joined, if not supplanted, in the popular imagination by two new meanings: running your charge card through a digital reader, and using your index finger to manipulate the touch screen of a smartphone. Swiping a charge card has become routine across much of the wired world, and once Apple rolled out the iPhone, swiping at smartphones quickly followed suit. Thanks to the commercial emergence of interactive touch screens, coupled with the popularity of apps, the smartphone is poised to become our most expansive and versatile payment technology yet, and the future of digital payment hinges on the swiping of a charge card giving way to swiping at a smartphone in order to complete transactions. Many smartphones and other touch screens for payment employ images of keys with numbers or letters with a layout virtually identical to those introduced a half century ago on touch-tone telephones.

The telephone keypad thus has an interesting connection to the swipe. It has enjoyed a long, successful career as an interface for retail transactions, from entering credit numbers over the phone to PINs in the checkout aisle. Debit cards and PINs outgrew the ATM and matured into an indispensable duo of payment technology. Few merchants in the United States today sell anything without accepting debit card payments. Many independent retailers have started using an attachment reader for smartphones and tablets in order to process transactions. Attachment readers like Square (manufactured exclusively for iPhones and iPads) are now commonplace at the cash registers of chains such as Starbucks as well as within more informal and independent settings such as farmers' markets and food trucks (see Mainwaring, this volume). Attachment readers are proving to be an affordable option for merchants big and small, the latest (and probably last) payment terminal organized around card swipes and keypads for entering PINs. Already PINs are increasingly entered onto a touch screen taking the shape of a keypad. Perhaps the most striking evidence of the keypad's resiliency is its widespread appearance on touch screens. With Apple Pay, if the swipe gesture is backgrounded, the "swipe fee" is accentuated—and there is a lesson here for contemporary payment.

Every time a customer pays with plastic, the card issuer charges the merchant an interchange fee, more commonly known as a swipe fee. In 2012, banks and credit card companies reaped over $15 billion in swipe fees. Swipe fees are now the second-highest operating expense for retail merchants, after labor costs. In the United States, swipe fees are the highest in the industrialized world. In Europe, by comparison, swipe fees cost merchants one-eighth of what they cost merchants in the United States. Swipe fee revenues have tripled in the United States in the past ten years, while the actual cost of processing a debit or credit card transaction continues to fall. In the wake of the 2008 financial crisis, swipe fees belatedly came in for federal regulation. The Durbin amendment to the Dodd-Frank Wall Street Reform and Consumer Protection Act of 2010 limited banks to charging merchants 21¢ per charge card transaction, plus 0.05 percent of each transaction. When swipe reform went into effect in October 2011, the average debit swipe fee on cards from covered banks dropped from 48¢ to 24¢ per transaction (the 21¢ fee plus 0.05 percent of the transaction, which is 3¢ on average). In 2012, these reforms saved consumers $5.8 billion and merchants

$2.6 billion.[1] But three years later, the new policies are still being appealed in federal court and each month banks in the United States continue to reap over $1 billion in revenue from swipe fees.

Meanwhile, some of the largest retailers in the world, including Macy's, Target, and Office Depot, have filed a swipe fee lawsuit against Visa and MasterCard, the duopoly controlling over 80 percent of the markets in debit and credit cards. The retail giants opted out of a $7 billion settlement that would have covered over seven million retailers nationwide (and been the largest settlement in US antitrust history). A total of fifteen retailers dropped out of the settlement and sued instead, claiming the settlement would have given the credit card duopoly too much freedom to raise swipe fee rates in the future.

The political economy of swiping involves some of the wealthiest companies in the world fighting over billions of dollars. The "swipe fee wars," as they have been called in some US media accounts, continue to be waged by some of the world's largest corporations and financial institutions, namely credit card companies and banks versus retail merchants, with everyday consumers caught in the cross fire. When companies like Walmart develop a transaction app and offer it to their customers, they promote its convenience and efficiency, and the savings in labor costs are presented as being passed along to shoppers as lower prices. But the appeal of transaction apps for retailers involves customers' satisfaction or loyalty far less than proprietary claims on their financial data. Payment apps help retailers capitalize (on) the wealth of personal information accessible via smartphones compared to the relative paucity of data that can be mined from the magnetic stripes of credit and debit cards, which are still the norm in the United States (unlike card chip technology already common in Europe and Canada).

As keys and push buttons give way to smooth surfaces as our leading digital interface, what can we learn about the future of touch screen payments from the history of the telephone keypad?

18.2 The most significant development in the keypad's fifty-year career as a payment technology was its adoption on ATMs

The telephone has been used to arrange payments and agree to prices since the days of operators and rotary dials, but the exchange of currency over the phone was not possible until monetary values could be viably, legally, and securely converted into ones and zeros. Digitization tethered retail payment to telecommunications infrastructure, inside stores as well as remotely. At the level of the interface, shoppers began using keypads to transact, first over the phone and now routinely at checkout terminals. The upgrade from rotary dials to keypads was an insidious one, garnering far less attention than the dial's replacement of operators. While dialing automated the everyday act of placing a telephone call, the keypad automated retail shopping by enabling push-button payments to become routine. Like the keypad before it, the touch screen is catalyzing an expansion of telephony, in the process remediating everyday transactions. When the routine mechanics of retail payment are reorganized, specifically around new technology from the cash register to the keypad, atomized tasks and responsibilities are often reassigned. For instance, once shoppers began to enter their PIN onto a keypad at the point of sale, the task of swiping one's charge card through a reader also quietly passed from cashiers to shoppers.

When Apple CEO Tim Cook unveiled the iWatch during his 2014 address to shareholders at their annual convention, he stood on stage and triumphantly shook his fists at the sky. But reactions to his presentation suggested that Cook's fortunes as heir to lionized founder Steve Jobs hinge on Apple's foray into digital payment. With first the iPod and then the iPhone, Jobs turned his annual keynote address to shareholders into the media launch of the year (after Apple broke away from the Consumer Electronics Show, the biannual industry-wide showcase for new tech); in 2014 on the big stage, Cook rolled out the latest iPhone as well as unveiling the iWatch, but the biggest splash of all promises to be made by Apple Pay. Digital systems for retail payment have proliferated, from PayPal to Square and more recently Stripe, Braintree, and Venmo, among others. None has generated the enthusiasm for digital payment that Apple Pay has. Cook's presentation immediately lifted the stocks of Visa and MasterCard because he announced that Apple had not built its own closed, proprietary network from scratch, which could have cut out credit card companies entirely; rather, the company was capitalizing on the eight hundred million credit and debit cards already on file with iTunes and Apple's App Store. Shares of Verifone, the leading manufacturer of debit card readers, shot up when Cook announced that Apple would open the system to outside hardware developers—anathema to the business model single-mindedly pursued (and perhaps perfected) by chair Jobs. Despite some glitches experienced by early adopters, more than two thousand banks nationwide alongside high-profile merchants like Whole Foods have installed Apple Pay (as have departments of the US federal government). Part of the unprecedented enthusiasm for Apple's introduction of a digital payment system—as opposed to Google's digital wallet, for instance—involves the relative security of its closed, proprietary networks, and expertise in wiring such systems into hardware like the iPhone. Yet most of the newfound hope in digital payment surrounds the sun that is Apple gadgets. The company's renowned consumer technologies serve as access points into its proprietary payment systems, within which Apple takes a cut from every transaction. Even before Apple Pay, purchase fees were Apple's second-largest revenue stream—second only to iPhone sales. Apple has long been involved with online payment, such as via the iTunes store, but Apple Pay promises to expand the company into an online retailer of staggering proportions.

18.3 The future of the keypad

Media coverage of hacks into the payment systems of store chains like Target and Home Depot (not to mention Chase bank) shone welcome light on the data collection and storage practices of retail corporations, alongside those of government agencies and telecom service providers. In response, Visa and MasterCard were able to impose a deadline of October 2015 for merchants to upgrade to smart card technology. The magnetic stripes found on the back of charge cards cannot alter encoded information, rendering it much more vulnerable to capture than on chip cards, which generate unique codes for each transaction. The upgrade to smart card readers will cost hundreds of dollars per machine, to the tune of $5 billion nationwide. A year before the deadline, less than 5 percent of the billions of charge cards in the United States were equipped with chips, and even fewer readers were ready for them. Shoppers in Europe and Canada have used smart cards for years; in Europe, the deadline for chip-literate readers was New Year's Day 2005. Finally, Visa and MasterCard garnered enough public and political support for the upgrade in the United States because of heightened concern about the security of shoppers' personal and financial information. The duopoly will spring for the new cards, but they are seeing to it that the costs of the new readers are borne by merchants. The cultural politics surrounding digital privacy and surveillance animate the political economy of the upgrade, from striped charge cards to smart cards. Most US shoppers wielding debit cards after the upgrade will still, at least initially, enter their PIN onto the keypad of a reader, but once the deadline passes, merchants without new readers will assume liability for fraud permitted by their outmoded machines. The upgrade to smart cards is providing the catalyst to reassign liability for fraud from card issuers to merchants. It's nonetheless worth remembering that liability for fraud will still only fall to the merchant, rather than the card issuer, once the cardholder has been proven innocent. As shoppers increasingly use their own smart phone as payment technology—instead of debit cards, PINs, and readers— what new tasks, responsibilities, and liabilities can we expect consumers to assume along the way? For starters, consumers will absorb the costs of purchasing and maintaining transaction technology, at least the end point access nodes of payment systems. And once consumers grow accustomed to using their own gadget to conduct retail purchases, more liability for fraud may fall to shoppers as well.

Three great leaps forward in the history of telephone interface have occurred: the dial, the keypad, and the touch screen. Each corresponds to a period of technological transformation in the telephone industry and society more broadly: the dial and automation, the keypad and digitization, and the touch screen and computerization. AT&T exercised monopoly control over telephony's automation and digitization, but the horizons of Apple's power are even vaster. Apple ushered in the computer era of telephony with the iPhone, and Apple Pay will help bring in the fourth great leap in telephone interface, not only when biometrics like fingerprints replace PINs for identification purposes, but when the provision of one's personal data gives way to them being pulled directly from your phone. The corresponding societal trend is less about financialization than datafication. Smartphones are becoming nodes of data transfer, including payment alongside personal information like account numbers and balances, purchase histories, and social media profiles. Before long, Apple Pay will be joined inside iPhones by iBeacon, an "indoor proximity system" allowing ads and promos to be pinged to shoppers' phones based on their location inside (or near) stores. Futuristic marketing fantasies aside, Apple is banking on iBeacon's low-power (and low-cost) transmitters to "close the loop" of e-commerce, whereby a purchase can be publicized (and bundled for sale, etc.) the instant it is made, via the same gadget. Apple's business model ultimately involves selling consumers both content and the platforms to run it, not unlike Sony did with CDs and Walkmans, but Sony never rang up anyone's purchase or claimed a slice of the fees collected therein. Soon

Apple will be able to take a cut of not only those transactions involving its apps or iTunes but also any transaction made with an iPhone. I, for one, was disappointed that Apple rejected iPay in favor of Apple Pay as the name for its new payments system, although obviously Pay Apple would have been more accurate.

Note

1. Robert Shapiro, "The Costs and Benefits of Half a Loaf: The Economic Effects of Recent Regulation of Debit Card Interchange Fees," Sonecon, 2013. https://nrf.com/sites/default/files/The_Costs_and_Benefits_of_Half_a_Loaf.pdf.

19.1 Topping up a mobile wallet

19

ETHER

RACHEL O'DWYER

In all forms of society there is one specific kind
of production which predominates over the rest,
whose relations thus assign rank and influence to
the others. It is a general illumination which bathes
all the other colors and modifies their particularity.
It is a particular ether which determines the specific
gravity of every being which has materialized
within it.

—Karl Marx, *Grundrisse*

Airtime Is Money

In June 2014, a Vietnamese gang of four made off with a bank loan worth VND35 billion (approximately US$1.5 million). The collateral used to secure the ill-gotten gains? Counterfeit airtime credit stubs worth approximately VND400 million. The plan was put into motion when the four purchased VND100,000 in credit vouchers from the Vietnam telecom company Mobi-Fone and proceeded to copy the codes imprinted on the vouchers onto fake cards at a Chinese factory. The forgeries were then smuggled back into Vietnam and used as assets in a fraudulent bank loan. The men involved worked as official MobiFone retailers, selling pay-as-you-go phone credit, and had previously borrowed from the bank using airtime credit stubs as collateral.

With pay-as-you-go plans, a user adds airtime to their phone to pay in advance for services such as talktime, texts, or data. Airtime takes the form of a unique numerical code printed onto a paper receipt, hidden beneath a silvered scratch card or added directly to a chip card associated with a user's mobile account. When this code is texted into the mobile network, it sets in motion a cryptographic process that guarantees the transfer of units of credit to a user's SIM card. But what happens if instead of texting the top-up code to your network and using the balance to catch up with a friend, text your partner and tell him you'll be late, or look up a weather report online, this code is texted to another phone number in the network, as a gift, to clear an outstanding balance on a bill or pay for public transport? These practices are now widespread. Today "airtime" is money. Phone credit tokens can be exchanged for cash or spent in retail stores in places such as Cote d'Ivoire, Egypt, Ghana, Uganda, and Nigeria. In Zimbabwe, where US banknotes have replaced the inflation-ravaged dollar and as many as eight different currencies are in circulation, airtime transfers, along with sweets and condoms, sometimes stand in for small transactions in the absence of available cash. Mobile money services like M-Pesa have formalized these transfers, and added features such as savings and microloans. New services have emerged that accept payment in airtime, such as the South African commuter service Air Taxi. Still other companies use airtime as a form of payment, with the mobile Internet company Jana paying its users in airtime to watch advertisements and test out the latest phone apps.

Airtime Is Ether

By 2016, a number of services existed that allowed people to pay for things using mobile phones, transforming our devices from communications toward payments and transactions. The crucial difference between mobile wallets or apps and something like airtime is that airtime doesn't just use the network as a channel for cash. Airtime distributors can create more airtime credits and adjust their exchange rate to state currencies at will.[1] It is a totally new kind of currency that directly transacts on the value of these ethereal channels. In fact, airtime trading is now so widespread that it has led to anxieties about the use of phone credit as an unregulated form of money. Carriers in such a scenario are not only providing communication. The money in question—what one commentator dubbed "ACUs" (airtime credit units)—is guaranteed by the underlying ether.[2]

Ether Is Money

Ether refers to the electromagnetic wavelengths that carry wireless and mobile signals, from radio to cellular networks, near field communications and the mobile Internet. Ether is invisible, made of the same "stuff" as the visible spectrum of color, only at

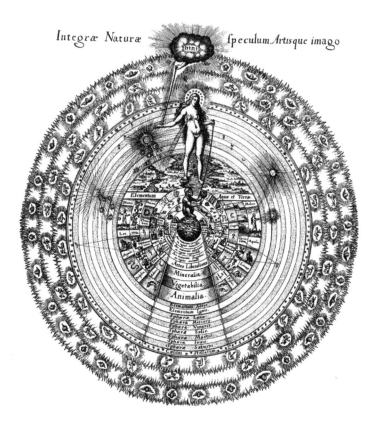

19.2 Robert Fludd's seventeenth-century engraving *Integrae naturae speculum artistique imago*

a dimension we can't see. Ether is immaterial. It consists of wave-like photons that have no materiality or "mass" but nonetheless interact with surrounding matter. So how does a continuum of frequencies become some "thing" that can be owned and controlled? What kinds of conceptual or imaginative leaps does this require?

In the nineteenth century, something called "the [a]ether" conceptually materialized the action and propagation of wireless signals. Scientists, inventors, and philosophers imagined an intangible dimension given over to the "flight of souls, voices without bodies and instantaneous presence at a distance."[3] While in reality these waves can travel in a vacuum, early accounts of action at a distance imagined an invisible "medium" to account for the propagation of light, heat, radio waves, and even telepathic communications. The properties of this ether were heavily debated, but most agreed that it was an invisible gas or fluid, a matter pushed to the limits of materiality—"much finer and rarer than air,"—that filled the whole of space.[4] The ether was a boundless medium that could act in any point in free space, collapsing time and distance, creating, as one of the founding inventors of radio, Oliver Lodge, put it, "links between mind and mind."[5] It was both a medium for communication and the element in which communications circulated. And as Lodge argued, "It may not be matter, but it must be some*thing*; there must be a connecting link of some kind, or the transference cannot occur."[6] Robert Fludd's seventeenth-century engraving *Integrae naturae speculum artistique imago* provides a visualization of this ethereal medium. The astrologer mapped the astrological spheres of the planets, stars, and matter of earth, and designated a rarefied middle region of the universe called the "middle spirit" or "ether" that acted as a conduit between the material and symbolic world.

The ether was a wireless medium for action at a distance that gained currency around the same time as railways and telegraphy, but early physicists also perceived it as a potential form of memory and storage. James Clerk Maxwell, responsible for the founding principles of electromagnetism, wrote of electromagnetic waves as a discovery that productively invigorated the waste spaces of the universe.[7] Taking this even further, his contemporaries Balfour Stewart and Peter Guthrie Tait would frame the ether as a kind of distributed "memory" device that worked against cosmic entropy, asserting that "the available energy of the visible universe [would] ultimately be appropriated by the ether."[8]

The Making of Ether

With the development of broadcasting in the early decades of the twentieth century, the concept of *ether* was used to imagine how wireless signals could be zoned, owned, and controlled by commercial interests. Previously wireless communications was a commons: nobody was thought to possess the ether, and anyone could transmit and receive messages provided they had access to the necessary equipment. But as time went on, and more and more people began to communicate wirelessly, and the Titanic sank in the Atlantic in a tangle of competing signals, economists argued for the need for licenses or property rights to govern the "avenues of the air."[9] The first radio acts declared the ether a scarce commodity in want of regulation.

Today this resource, which facilitates mobile communications, mobile payments, and of course, airtime, is divided up into bands of frequencies for specific purposes (such as cellular communications, satellite, radio navigation, and so on) and auctioned to the highest bidder. In many ways these auctions resemble higher-level transactions in airtime. And while mobile operators don't exactly trade ether like cash, a license represents "the true currency of wireless operators," counting for as much as 80 percent of a company's total net worth. But what exactly is at the heart of this value? Ether is currently amortized over a period of ten to forty years, and in accordance with a calculation known as megahertz pop (MHz pop), representing the human population in a license area multiplied by bandwidth. But the *ether* or *radio spectrum* is not a real thing or space; it is just a collection of potential values. A closer equivalent, as David Reed has contended, might be granting a corporation exclusive title to the color forest green in the state of New York.[10]

There Is No Ether

The existence of something called the ether was ultimately disproved in 1905 by Albert Einstein's theory of special relativity, which demonstrated that electromagnetic waves could travel without the aid of any medium at all. Invisible waves do not traverse some invisible medium or space known as *the ether*; in other words; they *are* the medium. Ether is just like any other thing—railway lines, fiber optics, and copper cables—that facilitates the circulation of stuff across time and space, then, but with the crucial difference that with ether, there is no "thing" as such. Nonetheless, long after the ether had fallen out of scientific favor, the concept continued to be used in journalism and policy.[11] There is no ether, but we continue to act and transact exactly as if there is. We sell it, trade it, and trespass in it. We grant exclusive rights to mobile network operators to sell us pay-as-you-go tokens to use bits of it. The System of National Accounts attributes billions of dollars' worth of value to it.[12] All of this collective belief points to the social contract at the heart

19.3 Artist Julian Oliver's *The Deep Sweep* (2015), an aerospace probe scanning the otherwise-out-of-reach signal space between land and stratosphere; intended for assembly and deployment by the public, it enables surveying and studying the vast, often-secretive world of signal in our skies of ethereal value—not that airtime credit stubs are guaranteed by a cosmic transmissions medium, but that this value is built on the communicative and cooperative capacities of a network of users. At the center of this value is nothing more than our communications with each other, and ether is a device to materialize these fleeting and transient connections. Ultimately it is this ethereal construction that lends a license its value, and underpins airtime as a medium of exchange or as collateral on a bank loan. Its materialization reveals a desire to control and "fix" social relations.

For media theorist Sybille Krämer, money is first and foremost a transmissions or media technology: it desubstantiates or dematerializes physical, lumpy property so it can circulate freely across time and space.[13] It mediates desires and wants while maintaining the distance between different economic actors. So is the ether an early kind of fiat currency that runs through networks of wireless signals and radios? It's true, we could think of ether as a kind of intangible money. But what ether really shows us is the slipperiness between money and materiality— one that cuts through debates around metallism versus chartalism. To say that money is ethereal is not the same as saying that it is made up, not real, or doesn't exist, a simple "coinage of the brain." Instead, it is to say that all money occupies a liminal space between the real and symbolic. Money is always on the verge of becoming something, and yet this very "thinginess" is also always frustrated. All money—physical coins or untethered data—is in the end a social contract, an ethereal substance that *is* and *is not* a thing.

Notes

1. Airtime Is Money," *Economist*, January 19, 2013, http://www.economist.com/news/finance-and-economics/21569744-use-pre-paid-mobile-phone-minutes-currency-airtime-money.

2. http://mondato.com/blog/airtime-as-currency.

3. John Durham Peters, *Speaking into the Air: A History of the Idea of Communication* (London: University of Chicago Press, 1999), 104.

4. Kenneth F. Schaffner, *Nineteenth-Century Aether Theories: The Commonwealth and International Library: Selected Readings in Physics* (Oxford: Pergamon Press, 1972), 7.

5. Oliver Lodge, *The Ether of Space* (London: Harper and Brothers, 1909), 99.

6. Ibid., 100.

7. "The vast interplanetary and interstellar regions will no longer be regarded as waste places in the universe, which the Creator has not seen fit to fill with the symbols of the manifold order of His kingdom." James Clerk Maxwell, *The Scientific Papers of James Clerk Maxwell* (New York: Dover, 1951), 708.

8. "All memory consists in an investiture of present resources in order to keep a hold upon the past. We have seen that this medium—this ether—has the power of transmitting motion from one part of the universe to another. A picture of the sun may be said to be travelling through space with an inconceivable velocity, and, in fact, continual photographs of all occurrences are thus produced and retained. A large portion of the energy of the universe may thus be said to be invested in such pictures." Balfour Stewart and Peter Guthrie Tait, *The Unseen Universe* (London: Macmillan and Company, 1879), 80.

9. The sinking of the Titanic was partly attributed to radio interference, which disrupted distress calls.

10. David Reed, quoted in David Weinberger, "Why Open Spectrum Matters: The End of the Broadcast Nation, 2003, http://apps.fcc.gov/ecfs/document/view;ECFSSESSION=W0sFWJVWD77rh5HXdgLyLJnBG6kVV1Zx2rPPk1nR2pmshdvWnbnd!310921635!-543955373?id=6513404739.

11. Herbert Hoover makes reference to the ether as a "public medium for public benefit" in his speech at the fourth national radio conference in 1925.

12. The System of National Accounts is the internationally agreed-on standard set of recommendations on how to compile measures of economic activity.

13. Sybille Krämer, *Medium, Messenger, Transmission: An Approach to Media Philosophy* (Amsterdam: Amsterdam University Press, 2015), 109.

20.1 Coin values at the Lancaster Trading House

20

SILVER

FINN BRUNTON

About a half hour north of the Mount Washington Hotel, in New Hampshire, and almost exactly seventy years to the day after it hosted the Bretton Woods Conference—the founding event of the postwar global monetary order—I was in a field in the White Mountains, using pieces of new-minted silver to purchase Wi-Fi access and a red Solo cup full of paleo cereal. After last night's thunderstorm, the air was still damp, lush, and almost visibly green. The Wi-Fi came over an antenna mounted on a trailer, connected to a mysterious 4G network on a virtual private network (VPN) with an exit node somewhere in Indonesia. "Paleo cereal" was, as it turned out, mostly almonds, pumpkin seeds, and coconut flakes. I received change in dimes—pre-1964 US dimes, with Franklin Delano Roosevelt's chin-up mug, looking toward the coin's milled edge, now a bit faint from decades of use.

It's a reasonable guess that there were more pre-1964 US coins in circulation in that field on that day than anywhere else in the world. This was a few days into PorcFest, a gathering named after porcupines, "peaceful and defensive" animals that want to be left alone—a friendly mammal version of the coiled rattlesnake on the "Don't Tread on Me" Gadsden flag. Any gathering of *n* libertarians has *n* + 1 definitions of what libertarianism means, but a few common components were generally shared: an ethos of leave-us-be armed self-defense (hence the porcupines) and free trade in open markets built on commodity money (hence the coins). As part of the larger Free State Project, PorcFest is loosely devoted to building a libertarian voting bloc in New Hampshire, but it also acts as a proof-of-concept space, a proving ground, for the performance of a libertarian society: to carry guns, take in technical lectures on free market economics and the work of Ayn Rand, drink raw milk, and engage in politically laden commerce.[1] It was that last part that concerned me—all those highly symbolic exchanges of literal value. Why were cryptocurrencies, like Bitcoin, being accepted by the same people taking payment in silver?

Silver! People in the community spent their free evenings with a bright light and sleeves of dimes and quarters from the bank, looking at the stacks for telltale signs of 90/10 coins—90 percent silver and 10 percent copper, as released by the US Mint from 1932 to 1964, and whose value as metal now considerably exceeds their face value as coins. They would pick these out and return the rest of the coins to the bank the next day, arbitraging the changes in the metallic content of coins over time for a small profit and collecting them to participate in this idiosyncratic payment system. Here in the field in New Hampshire, signs were posted: "1964 or before Silver Quarters" were accepted at a rate of one for a jar of Natalie's Raspberry Hemp Jam—or $3.75 in what they called "FRNs," that is, Federal Reserve Notes, also known as dollars and treated with bemused disdain. The old coins were prima facie evidence—literally, right on the face of the coin—that "value" was being produced out of nothing for the benefit of the state, making quarters from then and now commensurate. In Rand's *Atlas Shrugged*, the righteous buccaneer Ragnar Danneskjöld pays a character back for "the money that was taken from you by force" (that is, taxes and the like) with a bar of gold—"an objective value."[2] Mint master Bernard von NotHaus, who issued "Liberty Dollar" coins as well as gold and silver certificates from Hawaii and Idaho prior to his counterfeiting conviction in 2011, began with a manifesto, "To Know Value," denouncing the production of "money substitutes above the stored stock of real money"—"real" as in "a value based currency."[3]

Real, objective, true, universal, value-based: what it all means in this context is precious metal. From Rand (and her coincidental namesake, Krugerrands), to Liberty Dollars, the silver "deca" issued by Werner Stiefel's Atlantis initiative, and the "Nagriamel" coins (with the motto "Individual Rights for All") produced by the abortive project to turn Vanuatu into a libertarian enclave in 1977, the libertarian story could be told numismatically—a story of coins minted, exchanged, flourished, and seized.[4] This is a story into which Bitcoin does not seem to fit, and yet I could buy "Makin' Bacon Pancakes" at PorcFest with any one of four cryptocurrencies, along with silver and dollars ("and occasionally Hugs"). At a PorcFest picnic table, the "CURRENCY EXCHANGE BOARD," a gridded whiteboard with markers, had been set up to arrange transaction meetups: dollars for Dogecoin, Bitcoin for gold, and anything for silver. The silver in common use included those old coins, and a new lexicon of rounds and fractions stamped with the minters' marks of local operations. The most prominent mint present was the Suns of Liberty, based in Tamworth, about an hour and a half south. It sold fractional silver, quarter ounces in particular, just a little above the spot price. The silver pieces came in small black-velvet bags, whose popularity combined with the utilikilts, braided beards, shirtlessness, plant tinctures, flags and banners, and lack of working showers to give the general atmosphere of an extremely heavily armed Renaissance faire. The merchants under the boughs of the maples and blue spruce had small scales, calculators, and handwritten conversion charts for working out the effective payment value of different precious metals offered for gumbo and coffee, dry socks and jewelry, and a paperback collection of Lysander Spooner's essays. (Everything was still more or less denominated in dollars.) They had smartphones,

20.2 Makin' Bacon
Pancakes payment board

too, for checking the current bid-ask spread for precious metals
and doing transactions in Bitcoin.

The early Bitcoin mythos was rife with metallic comparisons,
from the original paper ("The steady addition of a constant of
amount [*sic*] of new coins is analogous to gold miners expend-
ing resources to add gold to circulation") to matters as seemingly
minor as Satoshi Nakamoto's birthday, entered once to set up a
profile—April 5, 1975.[5] April 5 is the anniversary of Roosevelt's

Executive Order 6102, forbidding the hoarding of monetary gold—an event that looms large in the libertarian imagination; in 1975, the prohibitions of the initial order were fully relaxed, and Americans could own and trade monetary gold again. Of course Bitcoin—and cryptocurrencies more generally—are, to a simple commonsense analysis, nothing whatsoever like precious metals. Bill Maurer, Taylor Nelms, and Lana Swartz have called this puzzle "digital metallism": a system that behaves like the credit theory of money while talking like the commodity theory—a commodity in the sense that money should be some intrinsically valuable substance, transformable into circuits, fillings, jewelry, photographic paper, colloidal silver, or whatever.[6] (One of the cars at PorcFest had vanity plates honoring Ludwig von Mises, the Austrian economist whose work argued that the basis for money-as-such as well as monetary policy lay in commodity money exchanged in a barter economy.) There are few things less intrinsically valuable than the scarce strings of letters and numbers required to solve the escalating difficulty of Bitcoin hashing problems, and the waste heat pouring off the boards of custom hashing-problem-solving chips that constitutes the work of "mining," and few arrangements that look more like a credit theory of money than a shared public ledger of abstract units of value generated according to a known rule to reward verifying transactions, and requiring buy-in, both literal and figurative, from a whole population of stakeholders. "The code and the labor are foregrounded because *they are practically all that Bitcoin enthusiasts ever talk about*": Maurer, Nelms, and Swartz hit the paradox precisely.[7] The same people for whom Bitcoin is the most objective currency—true in the way that the semiprime factors of prime numbers are true, universal and constant like gravity or the speed of light—worked on and talked about almost nothing else but how the currency existed by being worked on and talked about.

I feel affection for historical parallels, echoes, and ironies, and in coming to ask about the conflict between silver and cryptocurrency in practice, I was anticipating a mirror-image replay of the Bretton Woods Conference. This was ironic, because few people could express more eloquent and passionate contempt for what happened at that conference, for central bankers and international monetary planning, than the PorcFest participants, who wore "END THE FED" T-shirts and angrily discussed Keynesian monetary theory. A half hour south and seventy years ago, less than a month after D-day, John Maynard Keynes was among the seven-hundred-plus delegates rolling into New Hampshire to work out the economic architecture of the latter half of the century. In the run-up to Bretton Woods, Keynes had articulated one of the purest contemporary models of credit money backed by nothing but the utility of trade and balance of power between states, with a hypothetical currency called the *bancor*. (Keynes, not unaware of the clunkiness of the term, vacillated between *bancor* and *unitas* for the new unit; "both of them in my opinion are rotten bad names but we racked our brains without success to find a better.") The bancor would exist to "provide that money earned by selling goods to one country can be spent on purchasing the products of any other country. ... In English, a universal currency valid for trade transactions in all the world." You or I couldn't hold it; bancors would accumulate as a country exported, and deplete as it imported, lubricating the engine of commerce and preventing imbalances, backed by consensus and the value of exchange, not gold, because "there should be a supply of the money proportioned to the scale of the international trade which it has to carry."[8] The struggle of Bretton Woods concerned the arrangement of the postwar world, but it crucially circled around the guarantor of value that would underpin the global market: What it be US dollars, precious metals, international consensus, or the collective good of trade itself?

A few generations later, in the long aftermath of Bretton Woods, I asked, Why silver and Bitcoin? How did these merchants and individuals, who entirely opposed central bankers, global governance, and fiat currencies, believe in both of these seemingly dissimilar things? I was expecting from the conversations and debates "to know value" as one would know a fact about the world—that silver is the most electrically conductive of any element, for instance. Instead, I realized, this was a community where one "knew value" the way you knew it was hot outside

20.3 Currency Exchange Board

or you were among friends: bodily, interpersonal, sensory, social knowledge. What silver and Bitcoin shared weren't physical or ontological properties but rather ways that they could be evaluated and be *known*.

Cryptocurrencies in circulation are nothing more or less than records of creation, ownership, and transaction in the Blockchain ledger; their existence is constituted by the user-visible records of their existence. You could, in the words of one minter, "trust in yourself" to do the verification of what you hold when you hold silver. Silver is bodily: about palm feel, biting, body heat, and weight both on the scale and fingertip; about the look, gleam, and tarnish under different lights. (The minter argued against the use of security features on paper currency because they're a "distraction": they turn the verification into something someone else is in charge of, one step toward an abstract world of bancors and international order.) In the moment of payment, libertarian silver folds all the larger questions of money as such, of trade and value and currency, into the strange trust—with logic as circular as a coin—of only what we have before us in the moment of transaction between bodies in a damp New Hampshire field, as the minter said: "Well, silver is silver, and the weight is the weight."

Notes

1. See http://porcfest.com. In the context of the Free State Project, see https://freestateproject.org/events/porcfest.

2. Ayn Rand, *Atlas Shrugged* (1957; repr., New York: Signet, 1996), 253, 258.

3. United States v. NotHaus, 5:09-cr-00027-RLV-DCK, March 25, 2013, 9–10.

4. Monty Lindstrom, "Cult and Culture: American Dreams in Vanuatu," *Pacific Studies* 4, no. 2 (Spring 1981): 112.

5. Satoshi Nakamoto, Satoshi, "Bitcoin: A Peer-to-Peer Electronic Cash System," 2008, https://bitcoin.org/bitcoin.pdf. The site for the P2P Foundation requires a date of birth, which it then reflects in the profile's posted age; see http://p2pfoundation.ning.com/profile/SatoshiNakamoto?xg_source=activity. The cryptomarket analyst gwern (https://www.gwern.net), looking for the incrementing of the age through the archive of P2P Foundation, found that it fell on April 5, 1975.

6. Bill Maurer, Taylor C. Nelms, and Lana Swartz, "'When Perhaps the Real Problem Is Money Itself!': The Practical Materiality of Bitcoin," *Social Semiotics*, DOI:10.1080/10350330.2013.777594.

7. Ibid., 14.

8. John Maynard Keynes, "International Clearing Union," *House of Lords Debate*, May 18, 1943, 127:528–529.

Illustration Credits

p. v Photographs by Brian Ulaszewski (left) and Anthony Childs (right)

1.1 United States Patent 9262777 B2

1.2 Screenshot from http://giphy.com/gifs/amy-poehler-super-bowldongle -O5NQWAFrDwIZW

2.1 United States Social Security Administration

2.2 United States Social Security Administration

2.3 United States Department of the Treasury

3.1 Copyright © Phoebe A. Hearst Museum of Anthropology and the Regents of the University of California, Berkeley. Catalogue No. 1–1510.

3.2 Copyright © Phoebe A. Hearst Museum of Anthropology and the Regents of the University of California at Berkeley. Photography by Pliny E. Goddard, Hoopa, Humboldt County, 1901, Catalog No. 15–2947.

3.3 Copyright © Phoebe A. Hearst Museum of Anthropology and the Regents of the University of California at Berkeley. Catalog No. 1–1219.

4.1 CC0 Public Domain

4.2 Scan from author's personal collection

4.3 JPPI on MorgueFile.com, http://morguefile.com/archive/display/103074

5.1 Photo by the author

5.2 Photo by the author

6.1 Screenshot by the author

11.3 Microsoft Word Chinese character set

11.4 Institute for Money, Technology, and Financial Inclusion, University of California Irvine

11.5 Winchester City Council Museums, CC BY-SA 2.0

12.1 Photograph by the author

13.1 Friedberg Colonial, DE-76

13.2 Reproduced with the permission of the Library Company of Philadelphia

13.3 Reproduced with the permission of the Library Company of Philadelphia

13.4 Printed by Hall and Sellers, Philadelphia, 1776; reproduced with the permission of the Library Company of Philadelphia

14.1 Photograph by Michel Mailland

14.2 Photograph by Michel Mailland

14.3 Photograph by Jeff Jamison

14.4 Photograph by Jeff Jamison

14.5 Photograph by the author

14.6 Photograph by Jeff Jamison

14.7 Photograph by Michel Mailland

14.8 Photograph by the author

14.9 Photograph by the author

15.1 United States Patent Office

15.2 Photograph by Andrew Buxton

15.3 Photograph by Bernard Guyer

16.1 C. D. Ellithorpe, "Sidewalk Banking Apparatus," US patent 2700433

16.2 Courtesy of Ian Ormerod, NCR Fellowship

16.3 Courtesy of Corus Colors Record Centre

16.4 Courtesy of Jim Noll

16.5 Courtesy of Whybin/TBWA Group, Melbourne

16.6 Courtesy of Lloyds Banking Group Archives

17.1 National Numismatic Collection, National Museum of American History

17.2 National Numismatic Collection, National Museum of American History

17.3 National Numismatic Collection, National Museum of American History

17.4 National Numismatic Collection, National Museum of American History

17.5 National Numismatic Collection, National Museum of American History

17.6 Screenshot by Lana Swartz

17.7 From the author's personal collection

18.1 Photograph by Jina Valentine, Food Lion, Carrboro, NC

18.2 Photograph by Christina Dunbar-Hester

18.3 Screenshot by the author of his phone

19.1 Institute for Money, Technology, and Financial Inclusion, University of California, Irvine

19.2 United States National Library of Medicine

19.3 Image courtesy of the artist

20.1 Photograph by the author

20.2 Photograph by the author

20.3 Photograph by the author

Index

Note: Page numbers in *italics* indicate illustrations.

Shiba Inu dog meme, *52*, 54, 57–58, *59*, 61, 62, 67

Short-term versus long-term relationships, within sharing economy/ies, xix, 150, 151

Signature capture by signature pads (electronic signature pads). *See* Electronic signature pads (signature capture by signature pads); Signature/s on paper

Signature/s on paper

about, 129

Bitcoin digital signature versus, 128

credit card, 118–119, 122–123, 127, *128*

electronic signatures' relationship with, 118–119, 123–125

fraudulent, 123, 127–128

human signature as, 118

laws about, 123, 124–125

mortgage, 128–129

risks of unauthorized credit card, 123–124

Silver

certificates for, 251

as coins, 157, *248*, 250, 251

cryptocurrencies versus, 251, 252, *253*, 254, 255, 257

interchange among individuals and, 251, 256–257

libertarianism and, 250–251, 252, 257

tallies and, 135–136

value of exchange for, xx, 161, *248*, 251–252, *253*, 256, 257

Simmons, Matty, 86, 87

Siqueiros, David Alfaro, 108–109

Sirer, Emin Gün, 64–65

Smart cards. *See* Chip-embedded cards

Smartphones

data mining from, 226, 232

payments using, 30, 231–232

Square transactions, 8, 224–225

and swipe, the, 221

touch screens on, 224

Social Security program, xxiii, *12–13*, 14–17, *15*, *16*

Social status (equality/inequality), 89. *See also* Equality/inequality (distributional justice)

Socrates, 135